Margit Olle

Effective microorganisms influences vegetables and soybeans production

Margit Olle

Effective microorganisms influences vegetables and soybeans production

LAP LAMBERT Academic Publishing

Impressum / Imprint
Bibliografische Information der Deutschen Nationalbibliothek: Die Deutsche Nationalbibliothek verzeichnet diese Publikation in der Deutschen Nationalbibliografie; detaillierte bibliografische Daten sind im Internet über http://dnb.d-nb.de abrufbar.
Alle in diesem Buch genannten Marken und Produktnamen unterliegen warenzeichen-, marken- oder patentrechtlichem Schutz bzw. sind Warenzeichen oder eingetragene Warenzeichen der jeweiligen Inhaber. Die Wiedergabe von Marken, Produktnamen, Gebrauchsnamen, Handelsnamen, Warenbezeichnungen u.s.w. in diesem Werk berechtigt auch ohne besondere Kennzeichnung nicht zu der Annahme, dass solche Namen im Sinne der Warenzeichen- und Markenschutzgesetzgebung als frei zu betrachten wären und daher von jedermann benutzt werden dürften.

Bibliographic information published by the Deutsche Nationalbibliothek: The Deutsche Nationalbibliothek lists this publication in the Deutsche Nationalbibliografie; detailed bibliographic data are available in the Internet at http://dnb.d-nb.de.
Any brand names and product names mentioned in this book are subject to trademark, brand or patent protection and are trademarks or registered trademarks of their respective holders. The use of brand names, product names, common names, trade names, product descriptions etc. even without a particular marking in this work is in no way to be construed to mean that such names may be regarded as unrestricted in respect of trademark and brand protection legislation and could thus be used by anyone.

Coverbild / Cover image: www.ingimage.com

Verlag / Publisher:
LAP LAMBERT Academic Publishing
ist ein Imprint der / is a trademark of
OmniScriptum GmbH & Co. KG
Heinrich-Böcking-Str. 6-8, 66121 Saarbrücken, Deutschland / Germany
Email: info@lap-publishing.com

Herstellung: siehe letzte Seite /
Printed at: see last page
ISBN: 978-3-659-71844-1

Copyright © 2015 OmniScriptum GmbH & Co. KG
Alle Rechte vorbehalten. / All rights reserved. Saarbrücken 2015

Contents

	Page
Foreword	5
Paper 1	11
The effect of effective microorganisms on the yield and quality of tomatoes – a review	
Paper 2	23
The quality of tomato transplants influenced by effective microorganisms	
Paper 3	33
The influence of effective microorganisms on the growth and nitrate content of vegetable transplants	
Paper 4	45
The effect of effective microorganisms (EM) on the yield, storability and calcium content in Swede	
Paper 5	51
Effects of effective microorganisms on yield and quality of white cabbage	
Paper 6	63
Effective microorganisms effects on yield and quality of Chinese cabbage	
Paper 7	75
Influence of Effective Microorganisms on Soybean Seed Germination	

Foreword

Introduction to the book

This book gives an overview of the influence of effective microorganisms (EM) on the growth, development, yield and storage of greenhouse and field-grown vegetables.

The first paper is a review paper about the effect of EM on the yield and quality of tomatoes Effective microorganisms (EM) technology was first developed in the 1970ʻs. EM comprises a mixture of live cultures of microorganisms isolated from fertile soils in nature that are useful during crop production. Here, I review the effects of EM on tomato growth, yield, plant protection and chemical content. EM has been found to increase seed germination, vigour, early fruiting and the number of fruits in tomato. In most cases, it has also increased yield, probably due to increased photosynthesis. It is beneficial for pest management in tomato cultivation having pesticidal qualities. EM has reduced cucumber pickle-worm infection and is good against moth. EM can decrease phytophtora, blossom end rot and bacterial wilt incidence, but not bacterial, fungal and viral diseases of tomato. Applied with organic amendments it has enhanced weed growth in the first year which then declined significantly with time. It gave lower glycoalkaloid content in Bokashi-treated tomatoes. EM treatment increased vitamin C concentration and the content of saccharide, protein and amino acid in tomato. EM increased leaf N content and decreased leaf dry matter yield of tomatoes.

A literature search for the second paper revealed no publications on the effect of effective microorganisms (EM) on the growth parameters of tomato transplants. The aim of the study was to investigate how effective EM influence the growth parameters of tomato transplants. There were two experiments each with two treatments: 1. with EM; 2. without EM (control). Transplants with EM were significantly shorter and had a greater stem diameter than those without EM in both experiments. The stem diameter of transplants and the number of leaves per transplant did not differ significantly between treatments in either experiment. Transplants with EM produced fewer flowers than those without EM in one experiment but not in the other. We conclude that EM improve the quality of tomato transplants, as they remained more compact with a greater stem diameter.

In the third article, the rationale behind EM is based on the inoculation of soil with mixed cultures of beneficial microorganisms to create an environment more favourable for the growth and health of plants. The purpose of this investigation was to assess the influence of EM on the growth and nitrate content of cucumber and squash transplants. There were two treatments: 1. with EM; 2. without EM as control. In both experiments, cucumber, pumpkin and squash transplants grown with EM were significantly shorter and had thicker stems than those grown without EM. Nitrate content of transplants was lower in transplants grown with EM than in those grown without. In conclusion, EM improves the growth and reduces the nitrate content of cucumber, pumpkin and squash transplants.

Next, a field experiment with swedes. The aim of this investigation was to evaluate the effect of EM on the yield, storability and calcium content in swede.

The yield of swedes was highest in the treatment with EM with a yield increase of 26.7% greater than that of the treatment without EM. Storability was not influenced by EM. The Ca content of swedes after harvest was higher in the treatment with EM than in that without.

The aim of fifth study was to evaluate the influence of EM on the yield, storability and chemical content (nitrate, calcium, vitamin C, dry matter, monosaccharides and sugars) of white cabbage. There were two treatments: 1. with EM, 2. without EM as control. The yield of white cabbage was not influenced by EM treatment, although storage loss was lower with EM treatment. In 2013, after harvest, the content of monosaccharides after harvest was significantly lower and after storage significantly higher in the EM treatment. In 2012, the content of nitrate was significantly higher in the EM treatment both after harvest and after storage. Further, in 2012, the Ca content of white cabbage was also higher after storage in the EM treatment.

The sixth study evaluated the influence of EM on yield, storability and chemical content (nitrates, calcium, C-vitamin, dry matter, monosaccharides and sugars) of Chinese cabbage. Two treatments were compared: 1. with EM 2. without EM as control. EM did not influence yield but it decreased storage loss although not the dry matter content. The content of vitamin C was higher with EM treatment after 28 days of storage, but not after harvest. The content of monosaccharides, sugars and nitrates was not influenced by EM. With EM treatment, the content of calcium was higher after harvest, but not after storage. Thus, the main conclusions were that although EM treatment did not influence the yield of Chinese cabbage, it lowered storage loss, increased calcium content after harvest, and vitamin C content 28 days after storage. To conclude, EM generally had a positive influence on the growth, development, yield and storage, with greater yields and less storage losses. EM improve the quality of tomato, cucumber, squash and pumpkin transplants, as they remained more compact with a greater stem diameter.

The seventh paper was looking at the influence of EM on the soybean seed germination. As soybean has quite a long growing period, any effort made to shorten the growing period is beneficial. Therefore, it was important to conduct experiments to assess the effect of EM on soybean seed germination. The treatments included immersion of seeds for 60 minutes in activated EM solution 1:500 or in water (control). After immersion, the seeds were placed on moist filter paper in Petri dishes without drainage. There were significant differences among treatments in the germination percentage of soybean. The EM treatment showed the greatest number of germinated seeds.

A short introduction to senior researcher Margit Olle

Margit Olle (before marriage on 19.07.2003 Margit Kleemann) was born on 27.05.1971. She finished Secondary School No. 5 in Tartu with a **Silver medal**. She studied for her B. Sc. at the Estonian Agricultural University and finished in 1993 *Cum Laude*, passing all exams excellently. In 1995, she defended her M. Sc. at the same Institution. She took her doctoral studies in 1995 – 1999 in Norway at the Agricultural University of Norway and defended her Doctor Scientiarum degree on her 28th birthday on 27.05.1999 in Norway, Dr. Scient. degree was issued on the Estonian national public holiday Jaanipäev on 24.06.1999. Since 2008, she has worked as a senior researcher at the Estonian Crop Research Institute (former Jogeva Plant Breeding Institute). She speaks Estonian, English, Norwegian and some Russian. The main field of her work is scientific research in vegetable production on open land and in protected areas. She has published numerous review articles in addition to research articles in her field. During the years 2009 to 2014 she published six CC articles as first author. She has published a book: Kleemann, M. 2003. Vegetables for garden (in Estonian). 245 p. She organized (main responsible person) in 2001 (05.09 – 08.09) in castle Sagadi (Estonia) NJF seminar No. 329 entitled "New sights in vegetable production". She also organized (main responsible person – ISU chairperson) in 1997 (20.10 – 26.10) International Week in Aas at Agricultural University of Norway. Based on one CC review article she was invited as a speaker to "Korea – EU Photonics Business Forum" in Gwangju, South Korea on 06.12.13 to hold a lecture about "LED application in horticulture". Another CC article is being translated into Japanese. Since 2008, she has worked at the Estonian Crop Research Institute as a senior researcher on vegetable crops. Awards: 2015, Margit Olle; Margit Olle biography is in Who is Who in the World 2015 (32 Edition); 2015, Margit Olle; Margit Olle CV is in the 2000 outstanding intellectuals of the 21st Century, 9th Edition.

Private life: She is married to Vallo Olle and has a son Raimond Olle (born on 18.02.2004).

List of most important Current Content articles

Olle, M.; Tsahkna, A.; Tähtjärv, T.; Williams, I. (2014). Plant protection for organically grown potatoes – a review. Biological Agriculture & Horticulture: An International Journal for Sustainable Production Systems, xxx - xxx. [in press]

Olle, M.; Virǧle, A. (2013). The effects of light-emitting diode lighting on greenhouse plant growth and quality. Agricultural and Food Science, 22, 223 - 234.

Olle, M.; Williams, I. H. (2013). Effective microorganisms and their influence on vegetable production – a review. Journal of Horticultural Science & Biotechnology, 88(4), 380 - 386.

Olle, M.; Ngouajio, M.; Siomos, A. (2012). Vegetable quality and productivity as influenced by growing medium: a review. Ģemdirbystā=Agriculture, 99(4), 399 - 408.

Olle, M. (2012). Increase of leaf tipburn in chervil and lettuce by restricting volume of growing medium . Acta Agriculturæ Scandinavica, Section B - Soil & Plant Science, 62, 188 - 192.

Olle, M.; Bender, I. (2009). Causes and control of calcium deficiency disorders in vegetables: a review. Journal of Horticultural Science & Biotechnology, 84, 577 - 584.

Acknowledgements

In 2011, one of my colleagues **Margus Ess** introduced me to effective microorganisms (EM). He demonstrated in field experiments with corn crops that EM are very effective in increasing plant yields. I am very thankful for his considerable knowledge of this product and for my valuable discussions with him.

In 2012, I wrote a project proposal, which was approved and the study received financial support from the **Estonian Agricultural Registers and Information Board, with assistance from Jaagumäe Agro LLC and Estonian Crop Research Institute (former Jogeva Plant Breeding Institute).**

I have also had very good collaboration with the company **AgriPartner Ltd.**, whose manager **Argo Kukk**, has provided me with EM for my trials. **Argo Kukk** through his company **AgriPartner Ltd.** supported the visit of

New Zealand EM company staff in 2014. They visited the EM experiments at Estonian Crop Research Institute and some farms around Estonia using EM.

Special thanks to agronomist **Eve Somelar** for her correct work and her specialist knowledge of horticulture.

For English corrections and speciality discussions I am very thankful to **Ingrid Helvi Williams.**

Finally, a special thank-you to my mother **Aili Kleemann,** to my son **Raimond Olle** and to my husband **Vallo Olle** for always being there for me, for their understanding in every situation and for their support.

Paper 1

The effect of effective microorganisms on the yield and quality of tomatoes – a review

Unpublished paper

The effect of effective microorganisms on the yield and quality of tomatoes – a review

Abstract

Effective microorganisms (EM) technology was first developed in the 1970́s. EM comprises a mixture of live cultures of microorganisms isolated from fertile soils in nature that are useful during crop production. Here, I review the effects of EM on tomato growth, yield, plant protection and chemical content. EM has been found to increase seed germination, vigour, early fruiting and the number of fruits in tomato. In most cases, it has also increased yield, probably due to increased photosynthesis. It is beneficial for pest management in tomato cultivation having pesticidal qualities. EM has reduced cucumber pickle-worm infection and is good against moth. EM can decrease phytophtora, blossom end rot and bacterial wilt incidence, but not bacterial, fungal and viral diseases of tomato. Applied with organic amendments it has enhanced weed growth in the first year which then declined significantly with time. It gave lower glycoalkaloid content in Bokashi-treated tomatoes. EM treatment increased vitamin C concentration and the content of saccharide, protein and amino acid in tomato. EM increased leaf N content and decreased leaf dry matter yield of tomatoes.

Key words: chemical content, effective microorganisms, growth, plant protection, yield.

Introduction

Effective microorganisms (EM) technology was first developed in the 1970́s (Higa, 2012). EM comprises a mixture of live cultures of microorganisms isolated from fertile soils in nature that are useful during crop production (Mohan, 2008); these may include photosynthetic bacteria (e.g., *Rhodopseudomonas palustris*, *Rhodobacter sphaeroides*), lactobacilli (e.g.,

Lactobacillus plantarum, *L. casei*, and *Streptococcus lactis*), yeasts (e.g. *Saccharomyces* spp.), and Actinomycetes (*Streptomyces* spp.; Javaid, 2010).

The principle activity of EM is to increase the bio-diversity of soil microflora, thereby increasing crop yield. Photosynthetic bacteria, the major components of EM, are reported to work synergistically with other microorganisms to provide the nutritional requirements of the plant and to reduce disease (Condor et al., 2007). Subadiyasa (1997) described EM technology as one technique to support "natural farming". The rationale behind EM is based on inoculation of mixed cultures of beneficial microorganisms into soil to create an environment favourable for the growth and health of plants. EM interacts with the soil-plant ecosystem to suppress plant pathogens and agents of disease, to solubilise minerals, to conserve energy, to maintain soil microbial-ecological balance, to increase photosynthetic efficiency, and fix biological nitrogen (Subadiyasa, 1997).

Previous data from a randomised experiment showed statistically significant differences to indicate that EM increased seed germination and vigour in tomato (Siqueira *et al.*, 2012). Several authors have shown that the use of EM increases the yield of tomatoes (Marambe, Sangakkara, 1996; Zaenudin, 1993; Xu *et al.*, 2001).

Here, I review the effects of EM on tomato growth, yield, plant protection and chemical content.

EM effects on growth

Data from a randomised experiment showed statistically significant differences to indicate that EM increased seed germination and vigour in tomato (Siqueira *et al.*, 2012; Table 1). EM inoculation to both Bokashi and chicken manure increased photosynthesis of tomato plants (Xu *et al.*, 2001). The application of EM appeared to promote early fruiting in tomato (Ncube *et al.*, 2011).

Idris et al. (2008) found that EM treatment significantly increased plant height in tomato. In later stages (i.e. fruiting stages) this increase may have been through enhanced production of more primary branches and number of fruits. In contrast, EM may have a negative influence on the number of fruits (Ncube and Calistus, 2012); this could be because of nitrogen immobilization by EM, reducing its availability to the plants. The same research also revealed that the lower number of fruits associated with EM application resulted in improved average fruit weight of tomatoes grown in the greenhouse, possibly as a result of more assimilates being partitioned to the fewer fruits formed. However, other researchers have reported that EM treatment increases the number of fruits (Idris et al., 2008).

Conclusions: EM increases seed germination, vigour, early fruiting and the number of fruits in tomato.

EM effects on yield

Mohan (2008) evaluated the traditional Ayurvedic growth-promoters, Panchagavya and Amrit Pani, and compared them with 'Bokashi' made using EM technology (Table 2). He obtained higher yield from tomatoes treated with Bokashi, followed by Panchagavya. Similarly EM inoculation to both Bokashi and chicken manure increased yield (Xu *et al.*, 2001). EM treatments resulted in higher yield of tomatoes compared to the standard fertilizer treatment (Lee and Sung, 2001). In the same way, Escano (1996) showed that Bokashi and EM1 when used singly, or in combination with each other, or in combination with inorganic fertilizer, significantly increased mean fruit weight and the total marketable fruits harvested.

EM applied with a green manure (i.e., *Gliricidia* leaves) significantly increased tomato yields throughout the study; in the third year, the yields due to EM were comparable to those obtained with chemical fertilizer (Marambe and Sangakkara, 1996). In accordance, Zaenudin (1993) concluded that EM is needed in Indonesia because EM increases the production of tomatoes. The lower number of tomato fruits associated with EM application resulted in improved average fruit weight of tomatoes grown in the greenhouse, possibly as a result of more assimilates being partitioned to the fewer fruits formed (Ncube, Calistus, 2012). The highest yield of tomato resulted from weekly applications of EM at 1%; yield for that crop was 19.5% greater than the fertilized control (Wididana, Higa, 2013).

EM application in tomato production has sometimes resulted in a decrease in yield. Ncube et al. (2011) found that the application of EM alone resulted in a 26.9% decrease in fruit yield relative to the control while its application with compost resulted in a 23.2% decrease in fruit yield relative to treatment with compost. The authors pointed out that the apparent depressive effects of EM on tomato fruit yield could have been a result of a severe blight infestation on the crop which started in the EM-treated plots before rapidly spreading to the other treatments.

Conclusions: In most cases, EM has increased the yield of tomatoes, possibly by increasing photosynthesis.

EM effects on plant protection

The integrated results of Xu *et al.* (2012) suggest that the advantage of nitrogen metabolism in EM Bokashi-fertilized tomato plants accounted for the high phytophthora resistance (Table 3). The incidence of blossom-end rot of the EM treated plants was 35 % lower than that of control (Chen et al., 2012). Lee and Sung (2001) demonstrated that EM also decreased the incidence of blossom end rot and bacterial wilt (caused by [*Ralstonia solanacearum*]). Zaenudin (1993) concluded that EM is needed in Indonesia for pest management in tomato cultivation because of its pesticidal qualities.

Thaveechai *et al.* (1996) indicated that EM or SUTOJU was not effective against bacterial, fungal and viral diseases of tomatoes. Similarly Ncube et al. (2011) reported that EM did not protect against diseases, being totally ineffective in controlling the blights. They surmised that where EM has been found to have positive effects on tomato yield, the weather may not have been favorable for blight attack. In the Eastern Cape, South Africa where this experiment was carried out, the weather is at times very conducive to the development of blight and the results clearly indicate that EM may not be effective in controlling it.

Marambe and Sangakkara (1996) found that the application of organic amendments alone suppressed weed growth in tomato production, although the variation between years was not significant. EM applied with organic amendments enhanced weed growth the first year which then declined significantly with time.

Foliar applications of EM in FPE (fermented plant extract) and EM5 reduce organically grown cucumber pickle-worm infection (Condor et al., 2007). EM in FPE also proved best for moth control (Condor et al., 2007).

Conclusions: EM is beneficial for pest management in tomato cultivation because of its pesticidal qualities. It reduces cucumber pickle-worm infection and is good against moth. It can decrease phytophtora, blossom end rot and bacterial wilt incidence but does not protect against bacterial, fungal and viral diseases of tomato. EM applied with organic amendments enhanced weed growth the first year which then declined significantly with time.

EM effects on chemical content

Fruit chemical content

Mohan (2008) evaluated the traditional Ayurvedic growth-promoters, Panchagavya and Amrit Pani, which were compared with 'Bokashi' made using EM technology (Table 4). The results indicated lower glycoalkaloid content in Bokashi-treated tomatoes, followed by Panchagavya.

EM inoculation increased vitamin C concentration in fruit from all fertilization treatments (Xu, et al. 2001). EM treatment increased the contents of saccharide, protein and amino acid compared to control (Kim et al., 2013).

Leaf chemical content

The effect of EM on the N content of tomato leaves was investigated by Ncube at al. (2011). Combined applications of EM with amendments improved leaf N content compared to single application of the amendments.

EM alone increased leaf N content by 38.6% relative to the control. Application of EM + RF increased leaf N content by 15.1% relative to the mineral fertilizer treatment. These results confirm that EM interact with the soil-plant ecosystem to fix biological nitrogen (Subadiyasa, 1997).

A decrease in leaf dry matter yield with sole EM application was observed relative to the control (Ncube, Calistus, 2012). Therefore, Ncube and Calistus (2012) speculated that introduction of EM microbes into the soil could have set in short-term competition between the microbes and the plants for nutrients such as nitrogen in the limited pot soil volumes whose net effect was reduced plant growth. The suspected nutrient immobilization could also have been exacerbated by the introduction of carbon through molasses while applying EM to the soil. This could have stimulated indigenous microbial biomass pool activities in soil, causing N and P immobilization and reduced plant growth. This speculation is supported by the low N uptake observed in plots treated with EM.

Conclusions: Bokashi-treatment of tomatoes reduces their glycoalkaloid content. EM treatment increases vitamin C concentration, and the contents of saccharide, protein and amino acid in tomato fruit. EM increases leaf N content but decreases leaf dry matter yield of tomatoes.

References

Chen, J. N., Shao, X. H., Chang, T. T., Chen, L. H., Wang, W. N., Yin, C. 2012. Fermentation Production of EM Active Calcium and its Performance for the Prevention on Blossom-End Rot in Facility Tomato Cultivation. Advanced Materials Research, p. 610-613.

Condor, A. F., Gonzalez, P. and Lakre, C. 2007. Effective microorganisms: Myth or reality? The Peruvian Journal of Biology, 14: 315-319.

Escano, C. R. 1996. Experiences on EM technology in the Philippines. http://www.futuretechtoday.net/em/index2.htm, Accessed 13.11.2012.

Higa, T. 2012. Kyusei Nature Farming and Environmental Management Through Effective Microorganisms - The Past, Present and Future.

http://www.infrc.or.jp/english/KNF_Data_Base_Web/7th_Conf_KP_2.html Accessed 12.10.2012.

Idris, I. I., Yousif, M. T., Elkashif, M. E., Bakara, F. M. 2008. Response of tomato (*Lycopersicum esculentum* Mill.) to application of effective microorganisms. Gezira Journal of Agricultural Science, 6(1), North America, 6, oct. 2012. Available at: <http://journals.uofg.edu.sd/index.php/GJAS/article/view/4>. Accessed 06.04.2013

Javaid, A. 2010. Beneficial Microorganisms for Sustainable Agriculture. Sustainable Agriculture Reviews, 4: 347-369.

Kim, S. G., Lim, Y. D., Ryang, H. G. 2013. Yield and quality of vegetable as affected by effective microorganisms. www.emro-asia.com/data/7.pdf, Accessed 04.02.2013.

Lee, K. H., Sung, J. M. 2001. Improvement of Tomato Cultivation by Effective Microorganisms. Sixth International Conference on Kyusei Nature Farming. Proceedings of the conference on greater productivity and a cleaner environment through Kyusei Nature Farming, University of Pretoria, Pretoria, South Africa, 28-31 October, 1999, 2001 p. 75-79

Marambe, B. and Sangakkara, U. R. (1996). Effect of EM on weed populations, weed growth and tomato production in Kyusei nature farming. http://www.futuretechtoday.net/em/index2.htm. Accessed 11.12.2012.

Mohan, B. 2008. Evaluation of organic growth promoters on yield of dryland vegetable crops in India. Journal of Organic Systems, 3: 23-36.

Ncube, N. and Calistus, B. 2012. Effects of the integrated use of effective micro-organisms, compost and mineral fertilizer on greenhouse-grown tomato. African Journal of Plant Science, 6: 120-124.

Ncube, L., Minkeni, P. N. S. and Brutsch, O. 2011. Agronomic suitability of effective micro-organisms for tomato production. African Journal of Agricultural Research, 6: 650-654.

Siqueira, M. F. B., Sudré, C. P., Almeida, L. H., Pegorerl, A. P. R. and Akiba, F. 2012. Influence of Effective Microorganisms on Seed Germination and Plantlet Vigor of Selected Crops.

http://futuretechtoday.com/em/EMSeedGermPlantVigor.pdf   Accessed 14.11.2012

Subadiyasa, N.N. 1997. Effective microorganisms (EM) technology: its potential and prospect in Indonesia. Majalah Ilmiah Fakultas Pertanian Universitas Udayana, 16: 45-51.

Thaveechai, N., Boonwatana, N., Kamheangridthirong, T., Paradornuwat, A., Surin, P., Kositratana, W., Phawichit, S. and Buangiyapan, A. 1996) Efficacy evaluation of EM for plant disease control. Kasetsart Journal, 30: 67-76.

Wididana, G. N., Higa, T. 2013. Effect of EM on the Production Of Vegetable Crops in Indonesia. http://infrc.or.jp/english/KNF_Data_Base_Web/PDF%20KNF%20Conf%20Data/C4-4-123.pdf Accessed 02.03.2013.

Xu, H.L., Wang, R. and Mridha, M. A. U. 2001. Effects of Organic Fertilizers and a Microbial Inoculant on Leaf Photosynthesis and Fruit Yield and Quality of Tomato Plants. Journal of Crop Production, 3: 173-182.

Xu, H.L., Wang, R., Mridha, M. A. U. and Umemura, U. 2012. Phytophthora Resistance of Tomato Plants Grown with EM Bokashi. http://www.futuretechtoday.net/em/index2.htm Accessed 23.10.2013.

Zaenudin, S. 1993. Effective Microorganisms (EM4) Technology in Indonesia. http://www.futuretechtoday.net/em/index2.htm   Accessed 18.10.2012

1. The effect of EM on the growth of tomatoes.

Item	Effect	Reference
Seed germination	increased	Siqueira et al. 2012
Vigour	increased	Siqueira et al. 2012
Leaf photosynthesis	increased	Xu et al. 2001
Early fruiting	increased	Ncube et al. 2011
Plant height	increased	Idris et al. 2008
Number of fruits	increased	Idris et al. 2008

Table 2. The effect of EM on the yield of tomatoes

Solution	Effect	Reference
Bokashi	increased	Mohan, 2008
Bokashi	increased	Xu et al. 2001
EM	increased	Lee and Sung, 2001
Bokashi and EM1	increased	Escano, 1996
EM	increased	Marambe and Sangakkara, 1996
EM	increased	Zaenudin, 1993
EM	increased	Ncube and Calistus, 2012
EM	increased	Wididana and Higa, 2013
EM	decreased	Ncube et al. 2011

Table 3. The effect of EM on the plant protection of tomatoes

Item	Effect	Reference
Phytopthora	decreased	Xu et al. 2012
Blossom end rot	decreased	Chen et al. 2012; Lee and Sung, 2001
Bacterial wilt	decreased	Lee and Sung, 2001
Bacterial, fungal and viral diseases	not effective	Thaveechai et al. 1996
Blights	not effective	Ncube et al. 2011
Weed growth	enhanced first year, declined with time	Marambe and Sangakkara, 1996
Pickle-worm	decreased	Condor et al. 2007
Moth	decreased	Condor et al. 2007

Table 4. The effect of EM on the chemical content of tomatoes

Item	Effect	Reference
Glycoalkaloid in fruits	decreased	Mohan, 2008
Vitamin C in fruits	increased	Xu et al. 2001
Saccharide in fruits	increased	Kim et al. 2013
Protein in fruits	increased	Kim et al. 2013
Amino acid in fruits	increased	Kim et al. 2013
N in leaves	increased	Ncube et al. 2011
Dry matter in leaves	decreased	Ncube and Calistus, 2012

Paper 2

The quality of tomato transplants influenced by effective microorganisms

Original paper reference, published in Estonian language:
Olle, Margit (2014). Efektiivsete mikroorganismide mõju tomati istikute kasvule. Põllumajandusteaduselt tootjatele (100 - 103). Jõgeva 2014: Rebellis

The quality of tomato transplants influenced by effective microorganisms

Abstract

The aim of the study was to investigate how effective microorganisms (EM) influence the quality of tomato transplants. There were two experiments each with two treatments: 1. with EM; 2. without EM (control). Transplants with EM were significantly shorter and had a greater stem diameter than those without EM in both experiments. The width of transplants and the number of leaves per transplant did not differ significantly between treatments in either experiment. Transplants with EM produced fewer flowers than those without EM in one experiment but not in the other. We conclude that EM improve the quality of tomato transplants, as they remained more compact with a greater stem diameter.

Key words: effective microorganisms, height, stem diameter, tomato, transplants, quality

Introduction

Effective microorganisms (EM) comprise a mixture of live cultures of microorganisms isolated from fertile soils in nature that are useful during crop production (Olle 2013a). The principle activity of EM is to increase the bio-diversity of soil microþora, thereby increasing crop yield. Photosynthetic bacteria, the major components of EM, are reported to work synergistically with other microorganisms to provide the nutritional requirements of the plant and to reduce disease (Condor et al. 2007). Olle (2013b) described EM technology as one technique to support ñnatural farmingò. The rationale behind EM is based on inoculation of mixed cultures of beneficial microorganisms into soil to create an environment favourable for the growth and health of plants. EM interact with the soil-plant ecosystem to suppress plant pathogens and agents of disease, to solubilise minerals, to conserve energy, to maintain soil microbial-ecological balance, to increase photosynthetic efficiency, and to fix biological nitrogen (Olle 2013a).

Research has demonstrated that EM influence the germination, growth and yield of tomatoes. Siqueira et al. (2012) showed that they increase seed germination and vigour in tomato. Mohan (2008) compared the traditional Ayurvedic growth-promoters, Panchagavya and Amrit Pani, with 'Bokashi' made using EM technology. The Bokashi-treated tomatoes gave higher yield and a lower glycoalkaloid content, followed by Panchagavya. EM inoculation to both Bokashi and chicken manure increased photosynthesis and fruit yield of tomato plants (Xu et al. 2001). Field trials showed that Bokashi and EM1 when used singly, or in combination with each other, or in combination with inorganic fertilizer, increased mean tomato fruit weight and total marketable fruits harvested (Escano 1996). EM applied with a green manure (i.e., *Gliricidia* leaves) increased tomato yields throughout the study; in the third year, yields with EM were comparable to those obtained with chemical fertilizer (Marambe & Sangakkara, 1996). Zaenudin (1993) concluded that EM is needed in Indonesia because they increase tomato production. The application of EM appeared to promote early fruiting in tomato (Ncube et al. 2011). Although EM application to tomatoes in a greenhouse can result in fewer tomato fruits, these have a greater average fruit weight, probably because more assimilates are partitioned to the fewer fruits formed (Ncube & Calistus 2012).

Few investigations appear to have been carried out on the effects of EM on tomato transplant quality; only one is known to the authors of this article (Siqueira et al. 2012). In Estonia, tomato variety Valve transplants tend to become elongated so experiments were conducted to see whether use of EM would improve their quality.

Materials and Methods

Two experiments were conducted simultaneously in a heated greenhouse at the Jogeva Plant Breeding Institute during winter 2013, with the tomato variety Valve. There were two treatments: 1. with EM, 2. without EM (control), with four plants per treatment in each experiment.

Tomato seeds were sown on 16 January into seed trays and young plants were transplanted twice: firstly on 31 January at a spacing of 5 cm into larger trays and secondly, on 13 February, into individual pots (9 cm diameter). The substrate used was a peat-based mixture fertilised with PeatCare 11-25-24 2 kg m^{-3}, magnesium sulphate 0.5 kg m^{-3}, mixed with dolomite lime (7 kg m^{-3}).

Seeds for the treatment with EM were soaked in activated EM 1:500 solution for 30 minutes before sowing whereas those for the control were soaked in water only before sowing. They were sown on 16 January either in limed, fertilized and activated EM 1:500 solution treated peat (with EM) or in limed, fertilized and water treated peat (without EM). We used 2 L liquid per 12.5 L peat in both treatments. On 31 January, the larger trays were filled either with limed, fertilized and activated EM 1:500 solution treated peat (with EM) or with limed, fertilized and water treated peat (without EM). We used 2 L liquid per 12.5 L peat in both treatments. On 13 February, the pots were filled either with limed, fertilized and activated EM 1:1000 solution treated peat (with EM) or with limed, fertilized and water treated peat (without EM). We used 4 L liquid per 25 L peat in both treatments. From 20 February until 13 March, the transplants were watered weekly either with activated EM 1:1000 solution (with EM) or with water (without EM). We used 4 L liquid per 32 plants in both treatments.

The greenhouse lighting at plant level was approximately 12,000 lux from high pressure sodium lamps. The plants were additionally lit for 18 hours (23.00 ï 16.00h). A minimum day temperature of 20 °C and night temperature of 18 °C was maintained in the greenhouse.

On 18 March, the height, width, stem diameter and the number of leaves on the plants were recorded. Analyses of variance were carried out on the data obtained using Excel.

Results

Tomato transplants treated with EM were significantly shorter than those without EM in both experiments (Figure 1); in the first experiment, they were 21% shorter, and in the second, 22% shorter.

The width of transplants (Figure 2) and the number of leaves per transplant (Figure 3) did not differ significantly between with EM and without EM treatments in either experiment.

The stem diameter of transplants was significantly greater in those treated with EM than in those without EM in both experiments (Figure 4); in the first experiment, they were 33% greater and, in the second experiment, 22% greater.

There were significantly fewer flowers per transplant (by 24%) in the treatment with EM than that without EM in the first experiment but no significant difference between treatments in the second experiment (Figure 5).

Discussion

A search of the literature has not revealed any studies of the effect of EM on the quality of tomato transplants although some data has been published on the effect of EM on tomato yield.

In this study, we found that tomato transplants treated with EM were significantly shorter than untreated control transplants. This does not concur with Idris et al. (2008) who found that EM treatment significantly increased plant height; however, they measured plant height at fruiting whereas we measured the height of transplants. Perhaps later (i.e. in fruiting stages) EM increases plant height by promoting the production of more primary branches and fruits.

In one experiment, there were fewer flowers on EM treated than on untreated control plants. Similarly Ncube & Calistus (2012) reported that EM may have a negative influence on the number of fruits produced. They suggested that the EM may have immobilized nitrogen leaving less available to the plants. However, they also found that, although EM treated plants produced fewer fruits, their average weight was greater, possibly as a result of more assimilates being partitioned to the fewer fruits formed. In contrast to the study of Ncube and Calistus (2012), others have found that EM treatment increases the number of fruits (Idris et al. 2008).

EM improved the quality of tomato transplants, because they remained more compact with a greater stem diameter than untreated plants. Good tomato transplant quality results in higher yields, as reported Pavlovic et al. (1998). EM give a good start to tomato transplants because they solubilise minerals, including Ca, from the substrate. Ca influences many processes beneficially: plants with a higher Ca content have less disease, are attacked by fewer insects, and have better transport and storage qualities.

Conclusion

Effective microorganisms improved the quality of tomato transplants (variety Valve), by making them more compact and with a greater stem diameter.

Acknowledgements

The author would like to thank the Estonian Crop Research Institute for financial support to carry through this investigation.

References

Condor AF, Gonzalez P, Lakre C 2007. Effective microorganisms: Myth or reality? The Peruvian Journal of Biology 14: 315-319.
Escano CR 1996. Experiences on EM technology in the Philippines. http://www.futuretechtoday.net/em/index2.htm, Accessed 13.11.2012.
Idris II, Yousif MT, Elkashif ME, Bakara FM 2008. Response of tomato (*Lycopersicum esculentum* Mill.) to application of effective microorganisms. Gezira Journal of Agricultural Science, 6(1), North America, 6, oct. 2012.

Available at: <http://journals.uofg.edu.sd/index.php/GJAS/article/view/4>. Date accessed: 06 Apr. 2013

Marambe B, Sangakkara UR 1996. Effect of EM on weed populations, weed growth and tomato production in Kyusei nature farming. http://www.futuretechtoday.net/em/index2.htm. Accessed 11.12.2012.

Mohan B 2008. Evaluation of organic growth promoters on yield of dryland vegetable crops in India. Journal of Organic Systems 3: 23-36.

Ncube L, Calistus B 2012. Effects of the integrated use of effective microorganisms, compost and mineral fertilizer on greenhouse-grown tomato. African Journal of Plant Science 6(3): 120-124.

Ncube L, Minkeni PNS, Brutsch O 2011. Agronomic suitability of effective micro-organisms for tomato production. African Journal of Agricultural Research 6: 650-654.

Olle M 2013a. Efektiivsete mikroorganismide mõju köögiviljade saagile, kvaliteedile ja säilivusele. In: Aiandusfoorum 2013, 10 - 13.

Olle M 2013b. Efektiivsete mikroorganismide mõju kaalika saagile, keemilisele koostisele ja säilivusele. In: Agronoomia 2013, 174 - 178.

Pavlovic R, Petrovic S, Stevanovic D 1998. The influence of transplants quality on the yield of tomato grown in plastic house. Acta Hort 456: 81-86.

Siqueira MFB, Sudré CP, Almeida LH, Pegorerl APR, Akiba F 2012. Influence of Effective Microorganisms on Seed Germination and Plantlet Vigor of Selected Crops. http://futuretechtoday.com/em/EMSeedGermPlantVigor.pdf Accessed 14.11.2012

Zaenudin S 1993. Effective Microorganisms (EM4) Technology in Indonesia. http://www.futuretechtoday.net/em/index2.htm Accessed 18.10.2012

Xu HL, Wang R, Mridha MAU 2001. Effects of Organic Fertilizers and a Microbial Inoculant on Leaf Photosynthesis and Fruit Yield and Quality of Tomato Plants. Journal of Crop Production 3: 173-182.

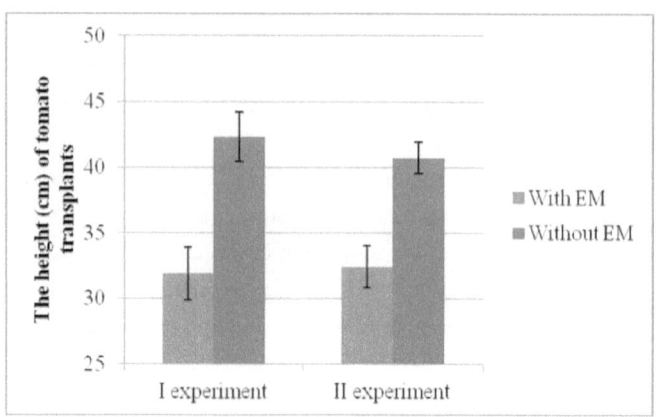

Figure 1. The height (cm) of tomato transplants (p=0.001 I experiment, p=0.00001 II experiment) on 7 March 2013.

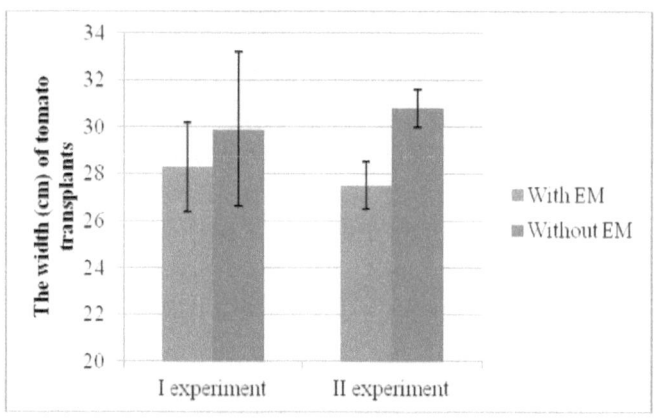

Figure 2. The width (cm) of tomato transplants (p=0.335 I experiment, p=0.002 II experiment) on 7 March 2013.

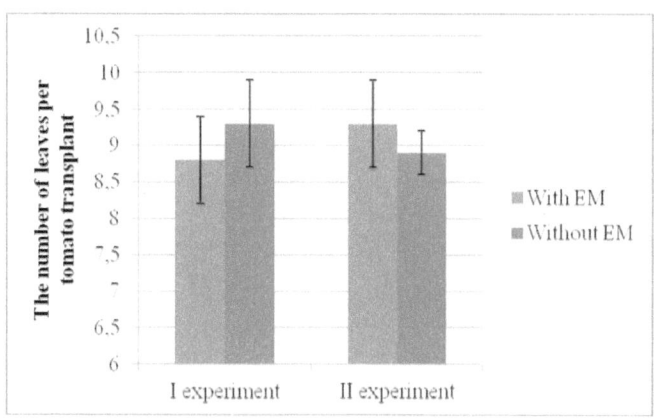

Figure 3. The number of leaves per tomato transplant (p=0.057 I experiment, p=0.097 II experiment) on 7 March 2013.

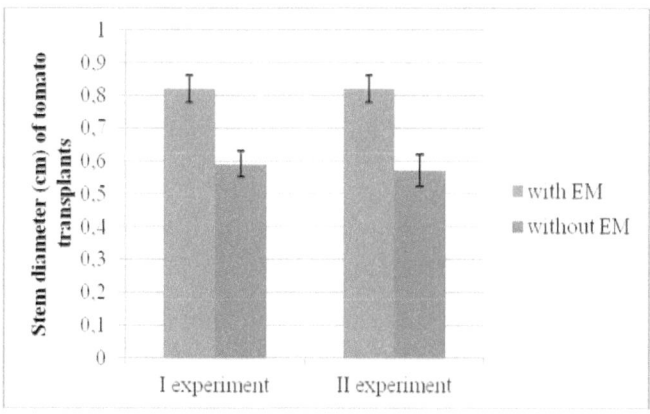

Figure 4. The stem diameter (cm) of tomato transplants (p=0.00001 I experiment, p=0.00001 II experiment) on 7 March 2013.

Figure 5. Photograph of tomato transplants (7 March 2013), on the left – tomato transplant treated with EM, on the right – tomato transplant not treated with EM.

Paper 3

The Influence of Effective Microorganisms on the Growth and Nitrate Content of Vegetable Transplants

Original paper reference.
Olle, M.; Williams, I. H. (2015). The Influence of Effective Microorganisms on the Growth and Nitrate Content of Vegetable Transplants. Journal of Advanced Agricultural Technologies, 2(1), 25 - 28.

The Influence of Effective Microorganisms on the Growth and Nitrate Content of Vegetable Transplants

Abstract— The rationale behind effective microorganisms (EM) is based on the inoculation of soil with mixed cultures of beneficial microorganisms to create an environment more favourable for the growth and health of plants. The purpose of this investigation was to assess the influence of EM on the growth and nitrate content of cucumber and squash transplants. There were two treatments: 1. with EM; 2. without EM – control. In both experiments, cucumber, pumpkin and squash transplants grown with EM were significantly shorter and had thicker stems than those grown without EM. Nitrate content of transplants was lower in transplants grown with EM than in those grown without. Conclusion: EM improves the growth and reduces the nitrate content of cucumber, pumpkin and squash transplants.

Index Terms— cucumber, effective microorganisms, height, nitrates, pumpkin, squash, stem diameter, transplants

Introduction

The rationale behind effective microorganism (EM) technology is based on the inoculation of mixed cultures of beneficial microorganisms into the soil to create an environment favourable for the growth and health of plants.

This technology was first developed in the 1970's [1]. Initially microbes from various ecosystems were isolated, then remixed. However, due to repeated lack of success, some microbes were eliminated, and simpler mixtures, comprising primarily lactic acid bacteria, photosynthetic bacteria, and yeast, maintained at pH 3.5. [1] were tested. Species used in an EM mixed culture of beneficial, naturally-occurring micro-organisms, may include the photosynthetic bacteria (e.g., *Rhodopseudomonas palustris*, *Rhodobacter sphaeroides*), lactobacilli (e.g., *Lactobacillus plantarum*, *L. casei*, and

Streptococcus lactis), yeasts (e.g. *Saccharomyces* spp.), and *Actinomycetes* (*Streptomyces* spp.) [2].

EM interact with the soil-plant ecosystem to suppress plant pathogens and agents of disease, to solubilise minerals, to conserve energy, to maintain the microbial-ecological balance of the soil, to increase photosynthetic efficiency, and to fix biological nitrogen [3].

Previous data from a randomized experiment showed statistically significant differences to indicate that EM increased seed germination and vigour in tomato [4]. Several authors have shown that the use of EM increases tomato yield [5]-[7].

It is well known that the quality of cucumber, pumpkin and squash transplants influences their final yields. The purpose of this investigation was to assess the influence of EM on the growth and quality of cucumber, pumpkin and squash transplants.

Materials and Methods

The experiments were carried out in spring 2014 in a heated glasshouse at the Estonian Crop Research Institute. The cucumber (*Cucumis sativus*) variety Landora F1, pumpkin (*Cucurbita maxima*) variety Atlantic Giant and the squash (*Cucurbita moschata*) variety Black Beauty were grown. There were two treatments: 1. with EM; 2. without EM (control).

Cucumber seeds were sown on 21 March into individual pots (9 cm diameter) and transplanted into 22 cm pots (24 April). Pumpkin and squash seeds were sown on 17 April into individual pots (14 cm diameter). The substrate for conventionally cultivated seedlings and transplants was a peat-based mixture fertilized with PeatCare 11-25-24 2 kg m^{-3}, magnesium sulphate 0.5 kg m^{-3}, mixed with dolomite lime (7 kg m^{-3}).

Seeds were soaked either in activated EM 1:500 solution (treatment 1) or in water (treatment 2) 30 minutes before sowing. Seeds were sown on in limed, fertilized and activated EM 1:500 solution treated peat (treatment 1) or in limed, fertilized and water treated peat (treatment 2). Once each week after sowing, plants were watered with either activated EM 1:500 solution (treatment 1) or with water (treatment 2) using 4 L liquid per 32 plants in each treatment.

Each treatment comprised 4 plants. The experiment had four replicates, grown concurrently.

The glasshouse lighting at plant level was approximately 12,000 lux from high pressure sodium lamps, lit for 18 hours (23.00 ï 16.00h) each day. Minimum day and night temperatures were 20 °C and 18 °C, respectively.

Plant height and stem diameter of cucumber were measured on 5 May and of pumpkin and squash on 12 May. Nitrate contents were determined in cucumber, pumpkin and squash transplants extracts by Fiastar 5000. Analyses of variance were carried out on the data obtained using programme Excel. Used signs: *** $p<0,001$; ** $p= 0,001 - 0, 01$; * $p= 0, 01 - 0, 05$; NS not significant, $p>0, 05$.

Results

EM treated cucumber plants were significantly shorter (by 16%) than control plants not treated with EM (Fig. 1).

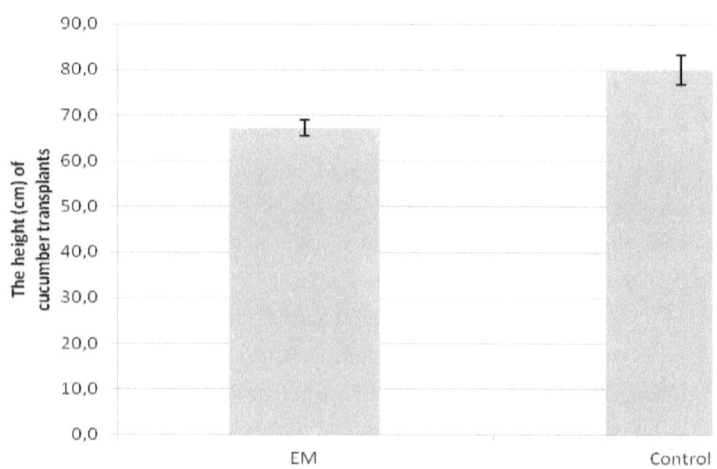

Figure 1. The height (cm) of cucumber transplants (***).

The stem diameter of cucumber transplants treated with EM was significantly greater (by 29%) than that of plants not treated with EM (Fig. 2).

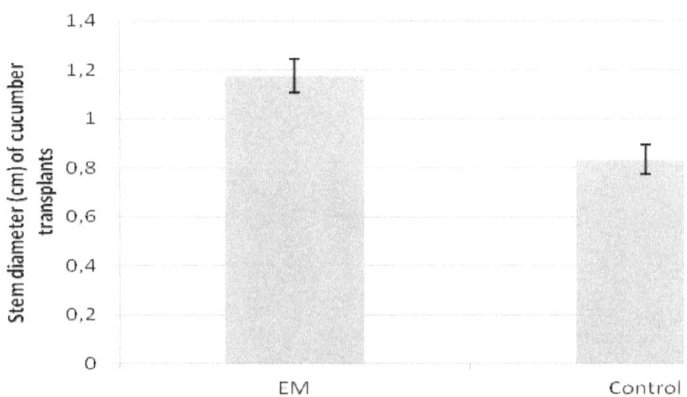

Figure 2. The stem diameter (cm) of cucumber transplants (***).

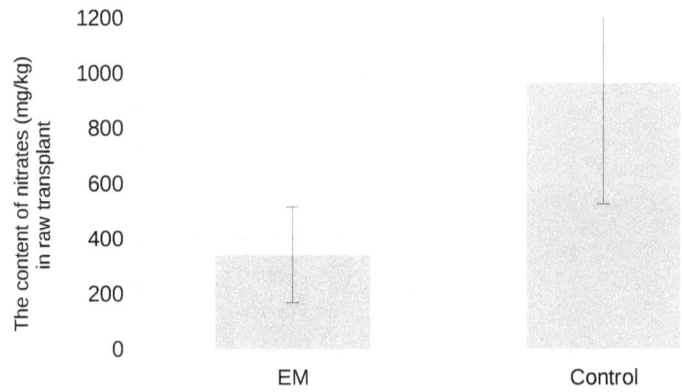

Figure 3. The content of nitrates (mg kg-1) of fresh cucumber transplants (*).

Nitrate content of fresh cucumber transplants treated with EM was significantly lower (65% less) than that of control transplants not treated with EM (Fig. 3).

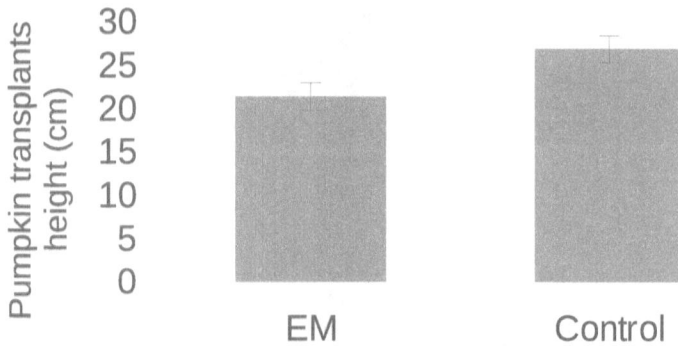

Figure 4. The height of pumpkin transplants (***).

Pumpkin transplants treated with EM were significantly shorter (by 20%) than those not treated with EM (Fig. 4).

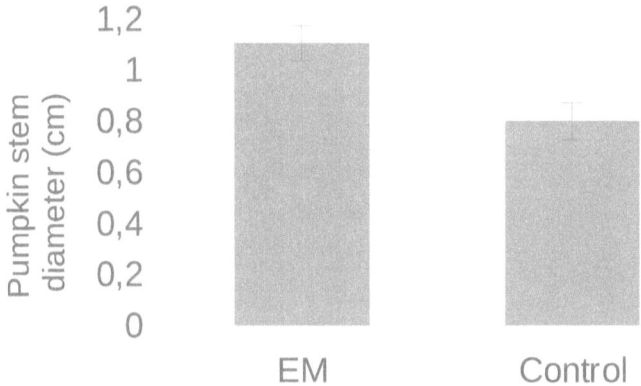

Figure 5. The stem diameter (cm) of pumpkin transplants (***).

The stem diameter of pumpkin transplants treated with EM was significantly larger (by 28%) than those not treated with EM (Fig. 5).

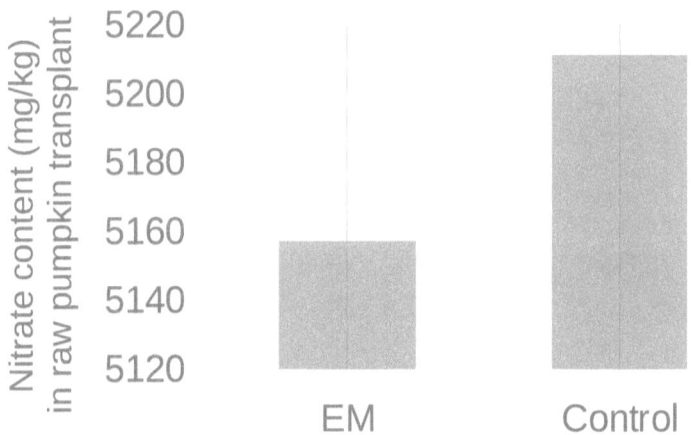

Figure 6. The content of nitrates (mg kg-1) of raw squash transplants (NS).

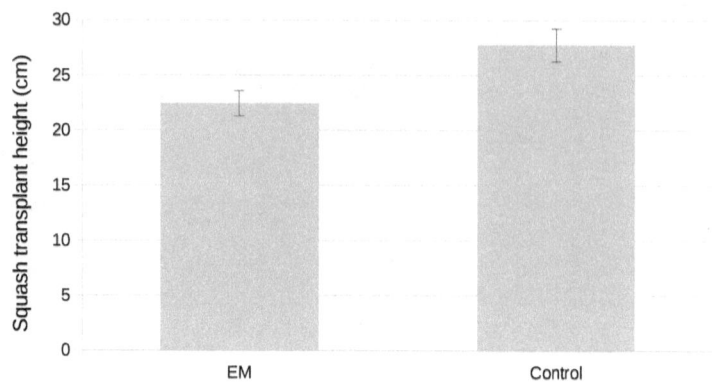
Figure 7. The height of squash transplants (***).

Nitrate content of fresh pumpkin transplants treated with EM tended to be less than that in those grown without EM, but not significantly different (Fig. 6).

Squash transplants treated with EM were significantly shorter (by 19%) than those not treated with EM (Fig. 7).

The stem diameter of squash transplants treated with EM was significantly larger (by 32%) than those not treated with EM (Fig. 8).

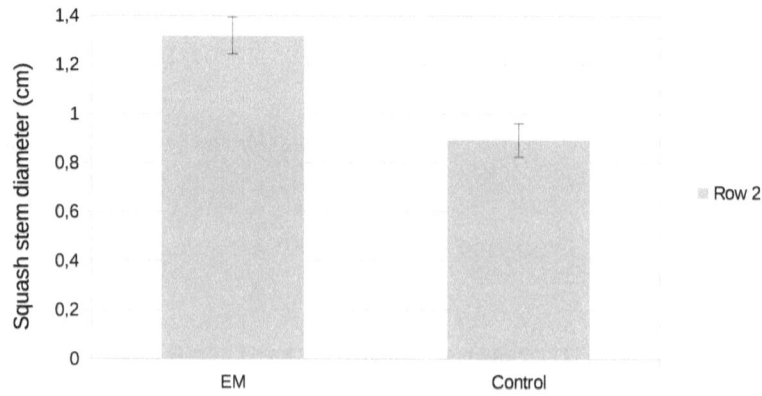
Figure 8. The stem diameter (cm) of squash transplants (***).

Nitrate content of fresh squash transplants treated with EM was significantly less (by 20%) than that in those grown without EM (Fig. 9).

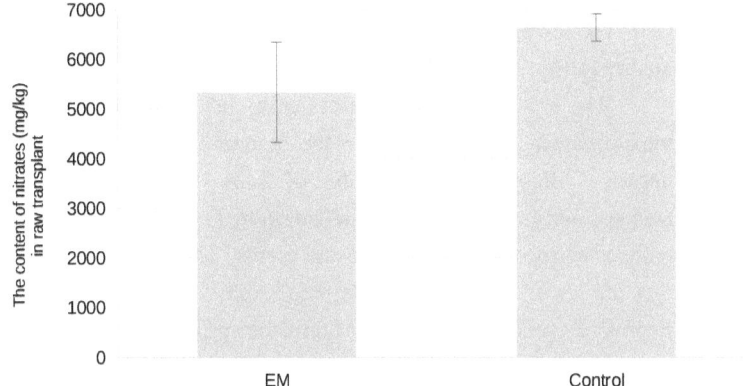

Figure 9. The content of nitrates (mg kg^{-1}) of raw squash transplants (*).

Discussion

Cucumber, pumpkin and squash transplants were significantly shorter when grown with EM than when grown without in both experiments. Similarly, EM treated tomato transplants are shorter than those grown without [8]. By contrast, treatment with EM has been found to increase plant height at fruiting [9]. Possibly in later stages (i.e. fruiting stages), EM would similarly increase the height of cucumber plants producing more fruits.

The stem diameter of cucumber, pumpkin and squash transplants was significantly larger in EM treated plants in both experiments than in EM untreated plants. The larger the stem diameter, the better the plant can obtain the nutrients from soil, especially Ca, which flows through the plants with the water flow of transpiration. Similar results have been obtained with tomato transplants [8]. In addition EM helps to solubilise minerals, including Ca, and the Ca content can increase in tomato plants [3]. This is desirable because a higher Ca content reduces the incidence of insect pests and diseases and improves fruit transport and storage qualities.

EM improved the quality of tomato transplants, because they stayed compact and stem diameter was greater. Good tomato transplant quality results in higher yields. This statement is supported by a study on tomatoes [10]. The lowest yield of tomatoes was noticed with the transplants of poorer quality [10].

The producers of EM claim that by using EM the number of leaves of tomato can increase. Data from the literature confirm that EM may have a negative influence on the number of fruits [11]. They pointed out that the reason could be nitrogen immobilization by EM, which could have resulted in reduced nitrogen availability to the plants. The same research also revealed that the lower number of fruits associated with EM application resulted in improved average fruit weight of tomatoes grown in the greenhouse, possibly as a result of more assimilates being partitioned to the few fruits formed. In contrast to previous investigation and research [11] some scientists found that EM treatment increases the number of fruits [12]. Mohan [13] evaluated the traditional Ayurvedic growth-promoters, Panchagavya and Amrit Pani, which were compared with 'Bokashi' made using EM technology. The results indicated higher yield and lower glycoalkaloid content in Bokashi-treated tomatoes, followed by Panchagavya. EM inoculation to both Bokashi and chicken manure increased photosynthesis and fruit yield of tomato plants [7]. For tomato, the generated data from initial field trials showed that Bokashi and EM1 when used singly, or in combination with each other, or in combination with inorganic fertilizer, significantly increased mean fruit weight over untreated control and increased the total marketable fruits harvested during the crop season [14]. EM applied with a green manure (i.e., *Gliricidia* leaves) significantly increased tomato yields throughout the study; in the third year, the yields due to EM were comparable to those obtained with chemical fertilizer [5]. In accordance, Zaenudin [6] concluded that EM is needed in Indonesia because it increases tomato production.

The content of nitrates was lower in EM treated transplants than in control transplants. If EM were found to reduce nitrate content also in cucumber, pumpkin and squash fruits, this would be a very desirable and important effect on their quality. The reason could be that nitrogen immobilization by EM, resulting in reduced nitrogen availability to the plants [11].

Acknowledgement

This research was financially supported by the Estonian Agricultural Registers and Information Board, the Estonian target financing project no. SF170057s09 and with the help of the Jaagumäe Agro OÜ and the Estonian Crop Research Institute.

References

1. T. Higa. (2012). Kyusei Nature arming and Environmental Management Through Effective Microorganisms - The Past, Present and Future. [Online]. Available:
http://www.infrc.or.jp/english/KNF_Data_Base_Web/7th_Conf_KP_2.html
2. A. Javaid, ñBeneficial microorganisms for sustainable agriculture,ò Sustainable Agriculture Reviews, vol. 4, pp. 347-369, 2010.
3. N. N. Subadiyasa, ñEffective microorganisms (EM) technology: Its potential and prospect in Indonesia,ò Majalah Ilmiah Fakultas Pertanian Universitas Udayana, vol. 16, pp. 45-51, 1997.
4. M. F. B. Siqueira, C. P. Sudré, L. H. Almeida, A. P. R. Pegorerl, and F. Akiba. (2012). Influence of Effective Microorganisms on Seed Germination and Plantlet Vigor of Selected Crops. [Online]. Available: http://futuretechtoday.com/em/EMSeedGermPlantVigor.pdf
5. B. Marambe and U. R. Sangakkara. (1996). Effect of EM on weed populations, weed growth and tomato production in Kyusei nature farming. [Online]. Available: http://www.futuretechtoday.net/em/index2.htm
6. S. Zaenudin. (1993). Effective Microorganisms (EM4) Technology in Indonesia. [Online]. Available: http://www.futuretechtoday.net/em/index2.htm

7. H. L. Xu, R. Wang, and M. A. U. Mridha, "Effects of organic fertilizers and a microbial inoculant on leaf photosynthesis and fruit yield and quality of tomato plants," Journal of Crop Production, vol. 3, pp. 173-182, 2001.

8. M. Olle, "Efektiivsete mikroorganismide mõju tomati istikute kasvule," Põllumaj and Usteaduselt Tootjatele, Jõgeva, Rebellis, pp. 100-103, 2014.

9. I. I. Idris, M. T. Yousif, M. E. Elkashif, and F. M. Bakara, "Response of tomato (*Lycopersicum esculentum* Mill.) to application of effective microorganisms," Gezira Journal of Agricultural Science. vol. 6, no. 1, October 2008.

10. R. Pavlovic, S. Petrovic, and D. D. Stevanovic, "The influence of transplants quality on the yield of tomato grown in plastic house," Acta Horticulturae, vol. 456, pp. 81-86, 1998.

11. L. Ncube and B. Calistus, "Effects of the integrated use of effective microorganisms, compost and mineral fertilizer on greenhouse-grown tomato," African Journal of Plant Science, vol. 6, pp. 120-124, 2012.

12. I. I. Idris, M. T. Yousif, M. E Elkashif, and F. M. Bakara. "Response of tomato (*Lycopersicum esculentum* Mill.) to application of effective microorganisms," Gezira Journal of Agricultural Science, North America, vol. 6, Oct. 2012.

13. B. Mohan, "Evaluation of organic growth promoters on yield of dryland vegetable crops in India," Journal of Organic Systems, vol. 3, pp. 23-36, 2008.

14. C. R. Escano. (1996). Experiences on EM technology in the Philippines. [Online]. Available: http://www.futuretechtoday.net/em/index2.htm

Paper 4

The effect of effective microorganisms (EM) on the yield, storability and calcium content in swede

Original Paper reference:
Olle, M. (2013). The Effect of Effective Microorganisms (Em) on the Yield, Storability and Calcium Content in Swede. In: International Plant Nutrition Colloquium and Boron Satellite Meeting Proceeding Book: International Plant Nutrition Colloquium and Boron Satellite Meeting, Istanbul/Turkey, 19-23 August 2013. Sabanci University, 714 - 715.

The effect of effective microorganisms (EM) on the yield, storability and calcium content in swede

Introduction

Effective microorganisms (EM) comprise a mixture of live cultures of microorganisms isolated from fertile soils in nature that are useful during crop production (Mohan, 2008). The principle activity of EM is to increase the biodiversity of soil microþora, thereby increasing crop yield. Photosynthetic bacteria, the major components of EM, are reported to work synergistically with other microorganisms to provide the nutritional requirements of the plant and to reduce disease (Condor et al., 2007). Subadiyasa (1997) described EM technology as one technique to support ñnatural farmingò. The rationale behind EM is based on inoculation of mixed cultures of beneficial microorganisms into soil to create an environment favourable for the growth and health of plants. EM interact with the soil-plant ecosystem to suppress plant pathogens and agents of disease, to solubilise minerals, to conserve energy, to maintain soil microbial-ecological balance, to increase photosynthetic efficiency, and to fix biological nitrogen (Subadiyasa, 1997). The aim of present investigation was to evaluate the effect of EM on the yield, storability and calcium content ofswede.

Methods

The experiment was carried out in 2012 in the fields of Jogeva Plant Breeding Institute in Estonia. A completely randomized design was used with 4 replications of each treatment. Two treatments were compared:
    1. with EM: one furrow (5.2m) in each plot was watered with (8 L 1:2000) activated EM suspension.
    2. without EM (control) : one furrow in each plot was watered with pure water (8L) only.

Treatments were applied on: 5 June, 19 June, 3 July and 17 July.

Each experimental plot was 5.2 x 2.8 m, i.e. 14.56 m^2. The rows were 65 cm wide with 10 cm interplant distance in rows. The swede variety Globus was grown. The main soil type in the experimental field was luvisol (KI); its chemical composition is given in table 1.

Table 1. Chemical composition of Swede experimental field soil.

pHKCl	Pmg/kg (AL)	Kmg/kg (AL)	Ca mg/kg	Mg mg/kg	Org. matter%
7.04	164.21	192.04	1193.75	84.19	2.82

The experimental field was ploughed in autumn 2011 and cultivated twice the following spring. Swede seeds were sown with a hand seeding machine on 3 May. Weeding was done by hand three times during the growing period.

Rainfall was low in the first decade of May and high in the second decade of August. The average temperature was less than 20 ^{0}C in July and only 15 ^{0}C during the second decade of July. However, the rainfall in May was sufficient for good seed germination and the seedlings grew well. The cool summer delayed development which was two weeks later than normal in mid- August.

The crop was harvested on 27 August. The swedes were stored for 28 days and the storage losses calculated. Determination of calcium was carried through by o-Cresolphthalein Complexone method (ISO 3696, in Kjeldahl Digest by Fiastar 5000). Analyses of variance were carried out on the data obtained using Excel. Average, p value and standard deviation were calculated.

Results and discussion

The yield of swedes was greater (by 26.7%) in the treatment with EM than without EM (Table 2). Storability was not influenced by EM (Table 2).

The Ca content of the swedes after harvest was higher in the treatment with EM than without EM. (Table 2).

Table 2. The yield (t/ha), yield loss (%) at 28 days after harvest and the content of Ca in dry matter (%) (p=0.912) in swedes with EM and without EM.

	Yield (t/ha)	Yield loss (%) after 28 days storage	Ca content (%)
With EM	30.96±2.54	5.75±0.97	0.52±0.01
Without EM	22.71±3.13	5.46±2	0.44±0.02
Probability level	p=0.023	p=0.805	p=0.001

There are no previous studies reported in the literature that investigate the effect of EM on swedes. The only root vegetable for comparison was radish. Yadav (2012) sprayed radish plants with EM suspension (1:500) and obtained a yield increase of 70.5%.

Weather conditions were unusually cold and wet in summer 2012. We gave EM suspension or pure water even if the soil was wet to investigate if there was an EM influence in these conditions. Results showed that EM influenced the yield positively.

The yield of swedes was highest in the treatment with EM. A possible explanation for the increase in yield is that EM helps to solubilise minerals and fix biological nitrogen the soil-plant ecosystem (Subadiyasa, 1997).

The Ca content of swedes after harvest was higher in those treated with EM.

Higher Ca content is beneficial for agricultural production, as:
- Å Disease incidence is reduced
- Å Insect pest infestation is reduced
- Å Transportability and storability of produce is enhanced

Conclusions

Treatment of swede with EM increased yield and calcium content.

Acknowledgements

The present research was financially supported by the Estonian Agricultural Registers and Information Board and with the help of the Jaagumäe Agro OÜ and Jogeva Plant Breeding Institute.

References

Mohan, B. (2008). Evaluation of organic growth promoters on yield of dryland vegetable crops in India. Journal of Organic Systems, 3: 23-36.
Condor, A. F., Gonzalez, P. and Lakre, C. (2007). Effective microorganisms: Myth or reality? The Peruvian Journal of Biology, 14: 315-319.
Subadiyasa, N.N. (1997). Effective microorganisms (EM) technology: its potential and prospect in Indonesia. Majalah Ilmiah Fakultas Pertanian Universitas Udayana, 16: 45-51.
Yadav, S. P. (2012). Performance of Effective Microorganisms (EM) on Growth and Yields of Selected Vegetables. http://www.futuretechtoday.com/em/background.htm (searched 28.12.2012)

Paper 5

Effects of effective microorganisms on yield and quality of white cabbage

Original paper reference:
Unpublished

Effects of effective microorganisms on yield and quality of white cabbage

Abstract

The aim of this research was to evaluate the influence of effective microorganisms (EM) on yield, storability and chemical content (nitrate, calcium, vitamin C, dry matter, monosaccharides and sugars) of white cabbage. There were two treatments: 1. ï with activated EM (watered with 1:2000 activated EM solution 4 times during the growth season), 2. ï without EM (watered with pure water 4 times during growth season), as control. Although the yield of white cabbage was not influenced by EM treatment, storage loss was lower. In 2013 after harvest, the content of monosaccharides was less but after storage it was higher with EM treatment than without. However, in 2012, the content of nitrate was significantly higher in the EM treatment after both harvest and after storage. In 2012, the Ca content was higher after storage in the EM treatment.

Additional keywords: storage loss

Introduction

Effective microorganisms (EM) comprise a mixture of live cultures of microorganisms isolated from fertile soils in nature that are useful during crop production (Mohan, 2008; Olle 2013a). The principle activity of EM is to increase the bio-diversity of soil microþora, which under some circumstances may increase crop yield. Photosynthetic bacteria, the major components of EM, are reported to work synergistically with other microorganisms to provide the nutritional requirements of the plant and to reduce disease (Condor et al., 2007). Subadiyasa (1997) described EM technology as one technique to support ñnatural farmingò. The rationale behind EM is based on inoculation of mixed cultures of beneficial microorganisms into soil to create an environment favourable for the growth and health of plants. EM interact with the soil-plant ecosystem to suppress plant pathogens and agents of disease, to solubilise minerals, to maintain soil microbial-ecological balance, and fix biological nitrogen (Subadiyasa 1997; Olle 2013b, Olle 2013c).

In previous experiments, the application of EM to Chinese cabbage appeared to promote root growth but not leaf and shoot development (In-Ho and Ji-Hwan, 2012). In head cabbage combined application of phosphorus fertiliser and EM enhanced vegetative and reproductive growth (Zahoor et al., 2003).

The aim of present investigation was to evaluate the effect of EM on the yield, chemical composition and storage losses in white cabbage. Ca content was determined as Ca is thought to affect storage life.

Materials and Methods

The experiments were carried out in 2012 and 2013 on the fields of Jogeva Plant Breeding Institute (since 1 July 2013 the Estonian Crop Research Institute) in Estonia. The experiment was a completely randomized design with 4 replicates:

There were 2 treatments:
1. with EM
2. without EM (control).

These were as follows:

At the time of planting (9 May 2012, 15 May 2013) either 1 L of 1:2000 activated EM suspension or 1L of pure water (control) was added to every plant in planting hole. There were three EM applications during the growing period, (in 2012 on 5 June, 19 June, 3 July and in 2013 on 31 May, 14 June, 28 June). Eight litres of 1:2000 activated EM suspension was also watered into one furrow in each plot (5.2 m). The treatment without EM got 8 L pure water.

The size of each experimental plot was 5.2 x 2.8 m i.e. 14.56 m^2, with 39 plants per plot and 156 plants per treatment. There were three rows per plot and each row was 65 cm wide. The distance between plants in rows was 40 cm. The variety of white cabbage grown was Reactor F1. The main soil type in the experimental field was a luvisol (KI). The chemical composition of the soil is given in Table 1.

Table 1. Chemical composition of white cabbage experimental field soil.

pHKCl	Pmg/kg (AL)	Kmg/kg (AL)	Ca mg/kg	Mg mg/kg	Org. matter %
7.04	164.21	192.04	1193.75	84.19	2.82

The experimental field was ploughed in autumn 2011 and 2012, and cultivated twice the following spring. White cabbage seeds were sown on 16 April 2012 and 9 April 2013 in the greenhouses of Jogeva Plant Breeding Institute. White cabbage transplants were planted on 9 May 2012 and 15 May 2013. Weeding was done by hand three times during the growing period. Cabbages were harvested on 10 August 2012 (due to cold and rainy summer) and on 1 July 2013 (normal weather).

In 2012, the summer was quite cold and the development was some weeks later than normal at the beginning of August. The summer in 2013 was close to many yearsôaverage climatic conditions in Estonia.

To determine the effects of the treatments on storage, the cabbages were stored at 10^0C and 95% relative humidity for 28 days and storage loss recorded. Determination of calcium was by the o-Cresolphthalein Complexone method (ISO 3696, in Kjeldahl Digest by Fiastar 5000). Nitrite and nitrate content was determined by Fiastar 5000. Dry matter loss of mass on drying was carried out by mixing the sample thoroughly with kieselguhr or a similar material, pre-drying the mixture, and finally drying for 2 h in a vacuum oven at $70\ ^0C \pm 1\ ^0C$ and approximately 6.6 kPa (66 mbar) pressure. Determination of vitamin C concentration was performed by titration (Redox Titration Using Iodate Solution). Sugars were determined colorimetrically, by the adjusted Bertrand method.

Analyses of variance were carried out on the data obtained using Excel and differences were deemed to be significant at $P<0.05$.

Results

The yield of white cabbage showed no statistical differences between EM treatment and control (Figure 1). Storage loss was smaller by 21% in 2012, and by 34% in 2013, in the EM treatment (Figure 2).

The content of dry matter and vitamin C did not differ significantly with treatment (Figures 3, 4). The content of monosaccharides did not differ statistically with treatment in 2012, but in 2013, after harvest it was significantly less and after storage significantly greater in EM treatment than without it (Figure 5).

The content of sugars did not differ significantly with treatment after harvest or after storage (Figure 6). The content of nitrate was significantly higher in the EM treatment after both harvest and storage in 2012, but not in 2013 (Figure 7). The Ca content was higher after storage in the EM treatment in 2012 (Figure 8), but there was no statistical difference between different treatments after harvest in 2012.

Discussion

Contrary to present investigation, in which the yield of white cabbage was not influenced by EM treatment, all previous studies on Chinese cabbage and head cabbage in the literature report only a positive effect of EM on yield. Chantal et al. (2010) reported that EM effectively increases cabbage yield. The reason could be improved photosynthesis. Yadav (2012) found that when EM foliar spray was applied at certain concentrations and time intervals crop yields were significantly increased. Foliar spray application of EM (1:500) at 15-day intervals was the most effective with 91.6% greater yield over control. The cabbage yield in plots with 1:1000 EM dilution sprayed at 45-day intervals was the lowest among EM sprayed plots but nevertheless was also 9.5% more than control. These results may demonstrate that EM solution has a positive impact on the growth and yield of cabbage, contrary to the results of the present investigation. Escano (1996) showed that plots treated with EM and Bokashi gave significantly higher cabbage yield compared with farmer's practice (NPK + chicken manure). Both seed pod/plant biomass and microbial density showed a maximum response to the EM Bokashi plus EM application (Nakano, 2007). Kim et al. (2012) showed that treatment with EM and EM-fermented compost in the greenhouse increased yield of Chinese cabbage in the field in autumn by 23.5 - 57.9%.

The reason for the reduction of storage loss in white cabbage after EM treatment may have been the higher Ca content after storage in these cabbages. More Ca: 1. decreases the physiologically active gibberellins, 2. this in turn increases Ca content, 3. inhibits permeability of cell membranes, 4. produces good turgor, 5. promotes strong cell walls, preventing leakage of cell liquids.

Summer 2012 was unusually cold and wet whereas summer 2013 was close to the long term average for Estonia. Hence, it is surprising that the data did not differ too much. The results showed that the yield of white cabbage was not influenced by EM treatment. Storage loss was lower in the EM treatment. The content of Ca was higher in the EM treatment after storage. The content of monosaccharides was in 2013 after harvest significantly higher in the control treatment and after storage it was significantly higher in the EM treatment. The content of nitrate was significantly higher in the EM treatment after harvest and after storage in 2012. The Ca content of white cabbage was higher after storage in the EM treatment in 2012.

The Ca content of white cabbage after storage was higher in those treated with EM. This is beneficial for agricultural production, as a higher Ca content:

- Reduces the incidence of disease
- Reduces infestation by insect pests
- Improves transportability and storability

Conclusions

The results showed that in white cabbage, treatment with EM did not influence the yield but reduced storage loss. In 2013, the content of monosaccharides after harvest was significantly lower but after storage was significantly higher in cabbages treated with EM than in those with no EM treatment. In 2012, cabbages treated with EM had a higher content of nitrate both after harvest and after storage, and a higher Ca content after storage, than cabbages not treated with EM.

Acknowledgements

This research was carried through with financial support from Estonian Agricultural Registers and Information Board and with the help of the Jaagumäe Agro OÜ and Estonian Crop Research Institute.

References

Chantal, K. Xiaohou, S., Weimu, W. and Basil, T. I. O. 2010. Effects of Effective Microorganisms on Yield and Quality of Vegetable Cabbage Comparatively to Nitrogen and Phosphorus Fertilizers. Pakistan Journal of Nutrition 9(11): 1039-1042.
Condor, A. F., Gonzalez, P. and Lakre, C. 2007. Effective microorganisms: Myth or reality? The Peruvian Journal of Biology 14: 315-319.
Escano, C. R. 1996. Experiences on EM Technology in the Philippines. http://www.futuretechtoday.net/em/index2.htm
In-Ho, H. and Ji-Hwan, K. 2012. Study on the Plant Growth Hormones. Retrieved on 20 December 2012 from http://www.futuretechtoday.com/em/study.htm
Kim, S. G., Lim, Y. D. and Ryang, H. G. 2012. Yield quality vegetables as affected by effective microorganisms. Retrieved on 13 November 2012 from http://www.syntropymalaysia.com/DownloadData/Agriculture/25_Yield_and_

Quality_of_vegetable_as_affected_by_Effective_%20Microorganisms_North_Korea.pdf

Mohan, B. 2008. Evaluation of organic growth promoters on yield of dryland vegetable crops in India. Journal of Organic Systems 3: 23-36.

Nakano, Y. 2007. Effects of Effective MicroorganismsÊ on the growth of *Brassica rapa*. Retrieved on 28 November 2012 from http://ebookbrowse.com/effects-of-effective-microorganisms-tm-on-the-growth-of-brassica-rapa-pdf-d18075139

Olle, M. and Williams, I. H. 2013a. Effective microorganisms and their influence on vegetable production ï a review. Journal of Horticultural Science & Biotechnology 88(4), 380 - 386.

Olle, M. 2013b. Efektiivsete mikroorganismide mõju köögiviljade saagile, kvaliteedile ja säilivusele. Aiandusfoorum 2013 10 - 13.

Olle, M. 2013c. Efektiivsete mikroorganismide mõju kaalika saagile, keemilisele koostisele ja säilivusele. Agronoomia 2013 174 - 178.

Subadiyasa, N.N. 1997. Effective microorganisms (EM) technology: its potential and prospect in Indonesia. Majalah Ilmiah Fakultas Pertanian Universitas Udayana 16: 45-51.

Zahoor, S., Ahmed, M.S. and Abbasi, N.A. 2003. Effect of phosphorus levels and effective microorganisms on seed production in cabbage (var. capitata). Sarhad Journal of Agriculture 19(2): 193-197.

Yadav, S. P. 2012. Performance of Effective Microorganisms (EM) on Growth and Yields of Selected Vegetables. Retrieved on 23 November 2012 from http://www.futuretechtoday.com/em/background.htm

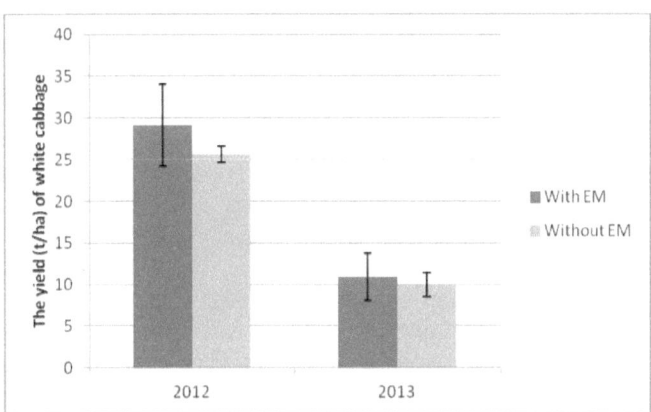

Figure 1. The yield of white cabbage (t/ha) (with and without EM) in 2012 and 2013; NS (not significant). The bars are the standard error of the mean.

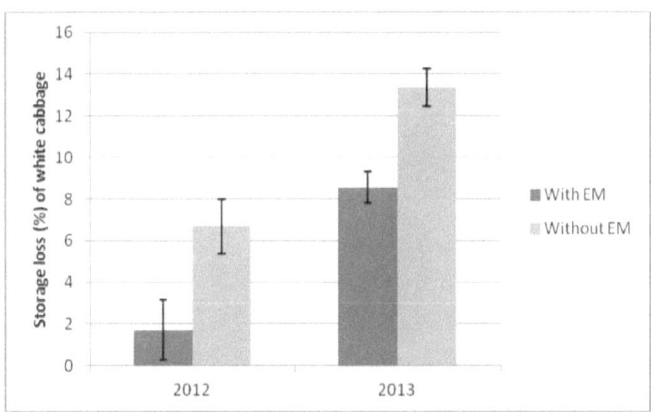

Figure 2. The white cabbage storage loss (%) (with and without EM) in 2012 and 2013. Treatment differences are significant ($P<0.05$) at 4 weeks of storage in 2012 and 2013. The bars are the standard error of the mean.

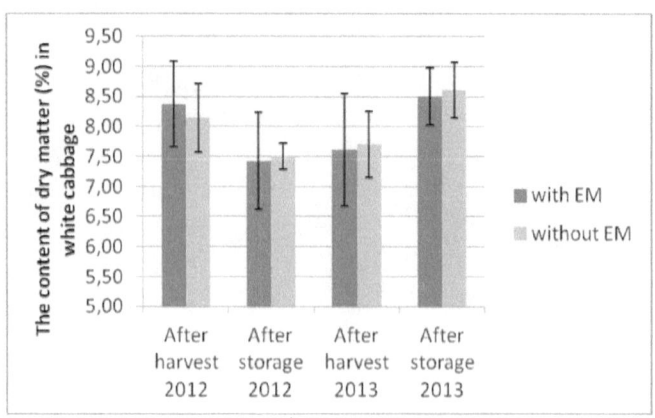

Figure 3. Dry matter (%) of white cabbage (with and without EM) in 2012 and 2013; NS. The bars are the standard error of the mean.

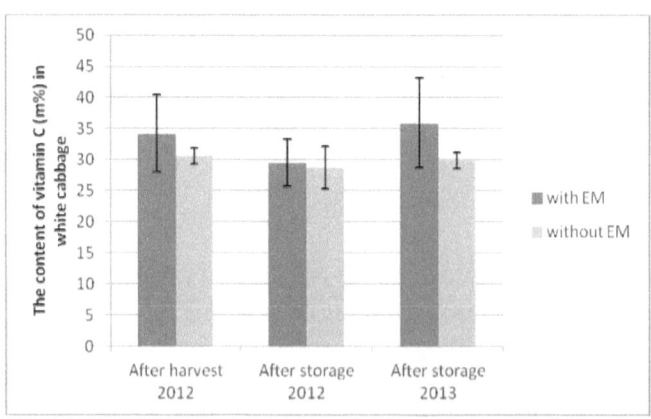

Figure 4. The content of vitamin C (m%) in white cabbage (with and without EM) in 2012 and 2013; NS. The bars are the standard error of the mean.

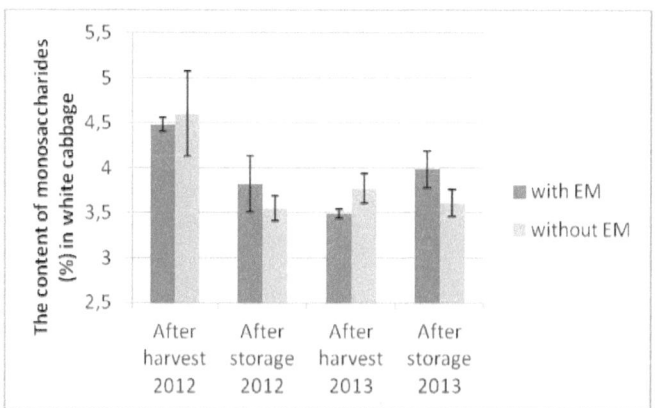

Figure 5. The content of monosaccharides (%) in white cabbage (with and without EM) in 2012 and 2013; after harvest 2012 and after storage 2012 NS, after harvest 2013 and storage 2013 significant ($P<0.05$). The bars are the standard error of the mean.

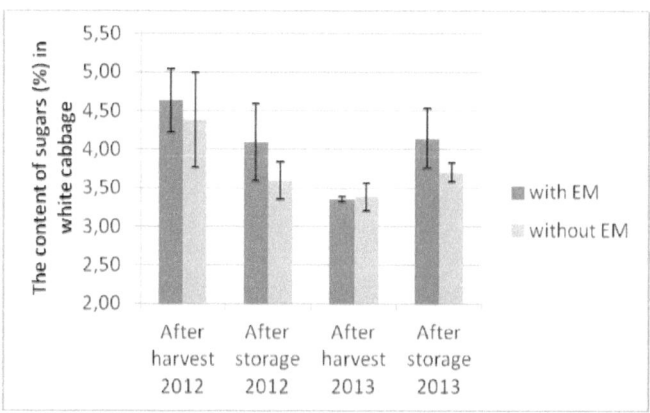

Figure 6. The content of sugars (%) in white cabbage (with and without EM) in 2012 and 2013; NS. The bars are the standard error of the mean.

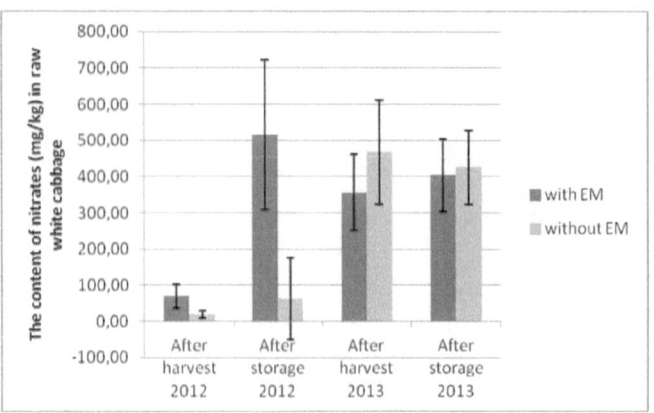

Figure 7. The content of nitrates (mg/kg) in raw white cabbage (with and without EM) in 2012 and 2013; after harvest 2012 and after storage 2012 significant (P<0.05), after harvest 2013 and after storage 2013 NS. The bars are the standard error of the mean.

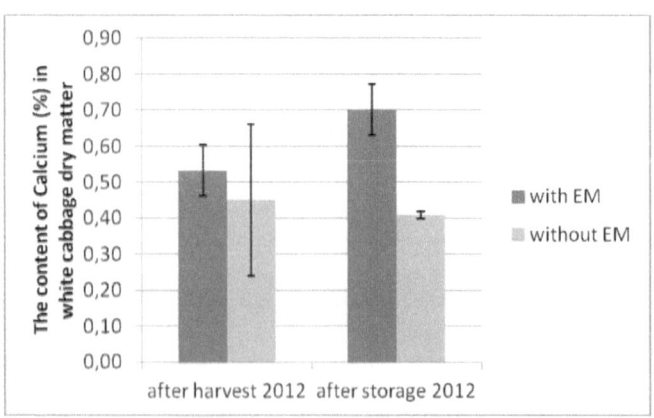

Figure 8. The content of calcium (%) in white cabbage dry matter (with and without EM) in 2012; after harvest 2012 NS, after storage 2012 significant (P<0.05). The bars are the standard error of the mean.

Paper 6

Effective microorganisms effects on yield and quality of Chinese cabbage

Original paper reference:
Unpublished

Effective microorganisms effects on yield and quality of Chinese cabbage

Abstract

The aim of this research was to evaluate the influence of effective microorganisms (EM) on yield, storability and chemical content (nitrates, calcium, C-vitamin, dry matter, monosaccharides and sugars) of Chinese cabbage. Two variants were used: 1 – with activated EM (watered with 1:2000 activated EM solution 4 times during the growing season), 2 – without EM (watered with pure water 4 times during the growing season) as control. The EM treatment did not influence yield or dry matter content but decreased storage loss. After harvest, EM treated and not treated cabbage did not differ in their content of C-vitamin but after 28 days of storage this had increased in the EM treated cabbage. The content of monosaccharides, sugars or nitrates was not influenced by EM, but the content of calcium was higher in the EM treated plants after harvest, although no differences remained after storage. The main conclusions were that treatment with EM did not affect the yield of Chinese cabbage but lowered storage loss, increased calcium content after harvest and vitamin C content at 28 days after storage.

Additional keywords: storability

Introduction

Effective microorganisms (EM) comprise a mixture of live cultures of microorganisms isolated from fertile soils in nature that are useful during crop production (Mohan, 2008; Olle 2013a). The principle activity of EM is to increase the bio-diversity of soil micropora, thereby increasing crop yield. Photosynthetic bacteria, the major components of EM, are reported to work synergistically with other microorganisms to provide the nutritional requirements of the plant and to reduce disease (Condor et al., 2007).

Subadiyasa (1997) described EM technology as one technique to support ñnatural farmingò. The rationale behind EM is based on inoculation of mixed cultures of beneficial microorganisms into soil to create an environment favourable for the growth and health of plants. EM interact with the soil-plant ecosystem to suppress plant pathogens and agents of disease, to solubilise minerals, to conserve energy, to maintain soil microbial-ecological balance, to increase photosynthetic efficiency, and fix biological nitrogen (Subadiyasa 1997; Olle 2013b, Olle 2013c).

In previous experiments with Chinese cabbage, application of EM appeared to promote root growth but not leaf and shoot development (In-Ho and Ji-Hwan, 2012). In head cabbage, combined application of phosphorus fertilizer and EM enhanced vegetative and reproductive growth (Zahoor et al., 2003).

The aim of the present investigation was to evaluate the effect of EM on the yield, chemical composition and storability in Chinese cabbage.

Materials and Methods

The experiment was carried out in 2012 on the fields of Jogeva Plant Breeding Institute (since 1 July 2013 the Estonian Crop Research Institute) in Estonia. A completely randomized experiment design was used with 4 replicates:

Two treatments were compared:
1. with EM
2. without EM (control).

They were treated as follows:

On 21 June (at planting) 1 L 1:2000 of activated EM suspension was added to every plant in its planting hole. Control plants received 1 L pure water instead of EM. Further treatments were applied on 3, 17 and 31 July; one furrow (5.2 m) in each plot was watered with (8 L 1:2000) activated EM suspension whereas the control treatment received 8 L pure water only.

Each experimental plot was 5.2 x 2.8 m, i.e. 14.56 m². Rows were 65 cm wide with 40 cm interplant distance within rows. The Chinese cabbage variety Suprin F1 was grown. The main soil type in the experimental field was luvisol (KI); its chemical composition is shown in table 1.

Table 1. Chemical composition of Chinese cabbage experimental field soil.

pHKCl	P mg/kg (AL)	K mg/kg (AL)	Ca mg/kg	Mg mg/kg	Org. stuff %
7.04	164.21	192.04	1193.75	84.19	2.82

In autumn 2011, the field was ploughed and in the following spring cultivated twice. Chinese cabbage seeds were sown on 1 June in the greenhouses of Jogeva Plant Breeding Institute and the seedlings transplanted on 21 June. Weeding was done by hand three times during the growing period. Plants were harvested on 13 August.

Rainfall was low in the first decade of May but high in the second decade of August. The average temperature for the whole of July was less than 20^0C and only 15^0C in the second decade. However, in May there was enough rainfall for seeds to germinate and seedlings to grow well. Because of the cold summer, development was two weeks later than normal at mid August.

For storability experiments, the harvested Chinese cabbages were stored at 10^0C and 95% relative humidity for 14 and 28 days and the storage loss was calculated. Determination of Calcium was carried through by the o-Cresolphthalein Complexone method (ISO 3696, in Kjeldahl Digest by Fiastar 5000). Nitrite and Nitrate contents were determined in vegetable extracts by Fiastar 5000. Dry matter determination was carried out by mixing the sample thoroughly with kieselguhr or a similar material, pre-drying the mixture, and finally drying for 2 h in vacuum oven at $70\ ^0C \pm 1\ ^0C$ and approximately 6.6 kPa (66 mbar) pressure. Determination of vitamin C concentration was performed by titration (Redox Titration Using Iodate Solution). Sugars were determined colorimetrically, by an adjusted Bertrand method.

Analyses of variance were carried out on the data obtained using Excel. Average, p value and standard deviation were calculated.

Results

The yield showed no significant differences between EM treated and EM untreated control Chinese cabbage plants (Figure 1). However, storage loss was influenced, being 34% less after 2 weeks and 21% less after 4 weeks in the EM treated than in the EM untreated control (Figure 2).

The content of dry matter after harvest and after 4 weeks storage did not differ with treatment (Figure 3), but after 2 weeks of storage dry matter content was less in the EM treated than untreated cabbage. The content of C-vitamin after harvest and after 2 weeks storage did not differ with treatment (Figure 4), but 4 weeks after storage was higher in the EM treated cabbage.

The content of monosaccharides (Figure 5), sugars (Figure 6) and nitrates (Figure 7) after harvest and after storage did not differ significantly between EM treated and EM untreated cabbage.

The content of calcium after harvest was higher in EM treated than EM untreated cabbage but not after 2 and 4 weeks storage (Figure 8).

Discussion

The yield of Chinese cabbage was not influenced by treatment with EM. This differs from previous published studies on both Chinese cabbage and head cabbage (e.g. Chantal et al, 2010) which report EM having a positive effect on yield, perhaps though improved photosynthesis. Yadav (2012) found that when EM foliar spray was applied at certain concentrations and time intervals crop yields were increased. Foliar spray application of EM (1:500) at 15-day intervals was the most effective with 91.6% more cabbage yield over control. The cabbage yield in plots with 1:1000 EM dilution sprayed at 45-day intervals was the lowest among EM sprayed plots but nevertheless was also 9.5% more than control. These results show that EM solution can have a positive impact on the growth and yield of cabbages, as opposed to the finding in the present investigation. Escano (1996) showed that plots treated with EM and Bokashi gave significantly higher cabbage yield compared with farmer's practice (NPK + chicken manure). Both seed pod/plant biomass and microbial density showed a maximum response to the EM Bokashi plus EM application (Nakano, 2007). Kim et al. (2012) showed that treatment with EM and EM-fermented compost in the greenhouse increased yield of Chinese cabbage in the field in autumn by 23.5 - 57.9%.

EM treatment reduced storage loss after 2 weeks in Chinese cabbage possibly because it promoted a higher Ca content at harvest. Increased Ca in plants: 1. decreases physiologically active gibberellins, 2. this in turn increases Ca content, 3. inhibits permeability of cell membranes, 4. resulting in good turgor, 5. stronger cell walls, preventing leakage of cell liquids.

Weather conditions were unusually cold and windy in summer 2012. EM suspension or pure water was applied even if the soil was wet to investigate any influence of EM in these conditions. Results showed that EM treatment reduced storage loss 2 weeks after harvest, increased C-vitamin content after 4 weeks storage, and increased the content of Ca in harvested yield.

The Ca content of Chinese cabbages after harvest was higher in those treated with EM. This is beneficial for agricultural production, as a greater Ca content:

- Å Reduces the incidence of diseases
- Å Reduces infestation by insect pests
- Å Improves transportability and storability

Conclusions

Although treatment with EM did not improve yield of Chinese cabbage, it decreased storage loss, and increased the content of Ca after harvest and of vitamin C 4 weeks after storage.

Aknowledgements

The present research was carried through with financial support from Estonian Agricultural Registers and Information Board and with the help of the Jaagumäe Agro OÜ and Jogeva Plant Breeding Institute.

References

Chantal, K. Xiaohou, S., Weimu, W. and Basil, T. I. O. 2010. Effects of Effective Microorganisms on Yield and Quality of Vegetable Cabbage Comparatively to Nitrogen and Phosphorus Fertilizers. Pakistan Journal of Nutrition 9(11): 1039-1042.

Condor, A. F., Gonzalez, P. and Lakre, C. 2007. Effective microorganisms: Myth or reality? The Peruvian Journal of Biology 14: 315-319.

Escano, C. R. 1996. Experiences on EM Technology in the Philippines. http://www.futuretechtoday.net/em/index2.htm

In-Ho, H. and Ji-Hwan, K. 2012. Study on the Plant Growth Hormones. Retrieved on 20 December 2012 from http://www.futuretechtoday.com/em/study.htm

Kim, S. G., Lim, Y. D. and Ryang, H. G. 2012. Yield quality vegetables as affected by effective microorganisms. Retrieved on 13 November 2012 from http://www.syntropymalaysia.com/DownloadData/Agriculture/25_Yield_and_ Quality_of_vegetable_as_affected_by_Effective_%20Microorganisms_North _Korea.pdf

Mohan, B. 2008. Evaluation of organic growth promoters on yield of dryland vegetable crops in India. Journal of Organic Systems 3: 23-36.

Nakano, Y. 2007. Effects of Effective MicroorganismsÊ on the growth of Brassica rapa. Retrieved on 28 November 2012 from http://ebookbrowse.com/effects-of-effective-microorganisms-tm-on-the-growth-of-brassica-rapa-pdf-d18075139

Olle, M. and Williams, I. H. 2013a. Effective microorganisms and their influence on vegetable production ï a review. Journal of Horticultural Science & Biotechnology 88(4), 380 - 386.

Olle, M. 2013b. Efektiivsete mikroorganismide mõju köögiviljade saagile, kvaliteedile ja säilivusele. Aiandusfoorum 2013 10 - 13.

Olle, M. 2013c. Efektiivsete mikroorganismide mõju kaalika saagile, keemilisele koostisele ja säilivusele. Agronoomia 2013 174 - 178.

Subadiyasa, N.N. 1997. Effective microorganisms (EM) technology: its potential and prospect in Indonesia. Majalah Ilmiah Fakultas Pertanian Universitas Udayana 16: 45-51.

Zahoor, S., Ahmed, M.S. and Abbasi, N.A. 2003. Effect of phosphorus levels and effective microorganisms on seed production in cabbage (var. capitata). Sarhad Journal of Agriculture 19(2): 193-197.

Yadav, S. P. 2012. Performance of Effective Microorganisms (EM) on Growth and Yields of Selected Vegetables. Retrieved on 23 November 2012 from http://www.futuretechtoday.com/em/background.htm

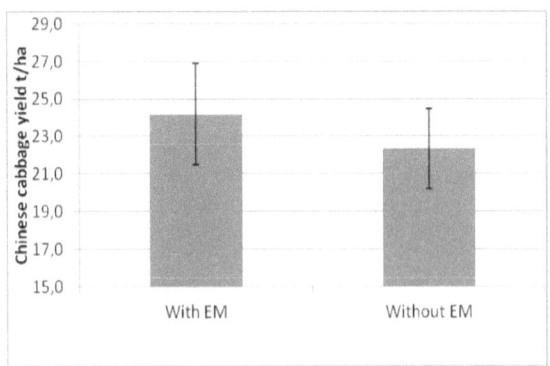

Figure 1. The yield of Chinese cabbage (t/ha) (with and without EM); NS (not significant).

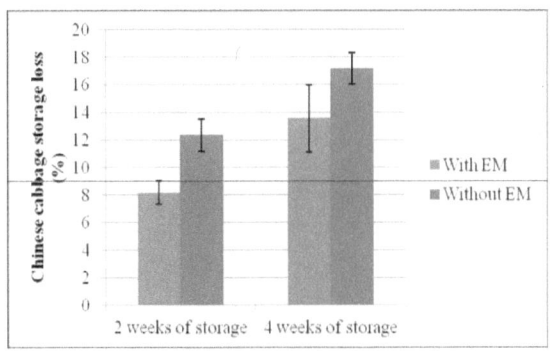

Figure 2. The Chinese cabbage storage loss (%) (with and without EM); (2 weeks of storage ï 0.05%), (4 weeks of storage ï 0.05%).

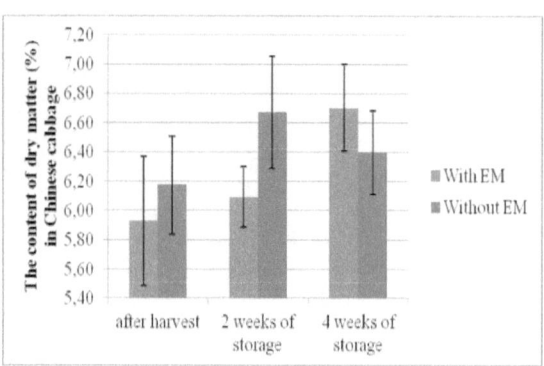

Figure 3. The content of dry matter (%) in Chinese cabbage (with and witout EM); after harvest NS, 2 weeks of storage 0.05%, 4 weeks of storage NS.

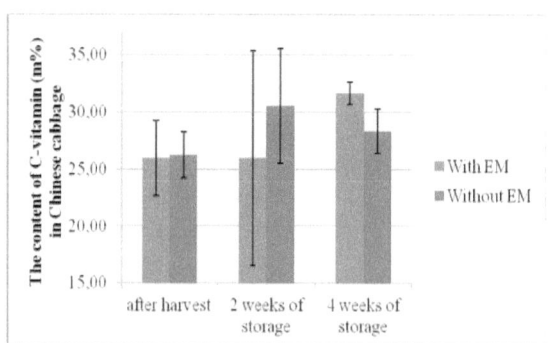

Figure 4. The content of C-vitamin (m%) in Chinese cabbage (with and without EM); after harvest NS, 2 weeks of storage NS, 4 weeks of storage 0.05%.

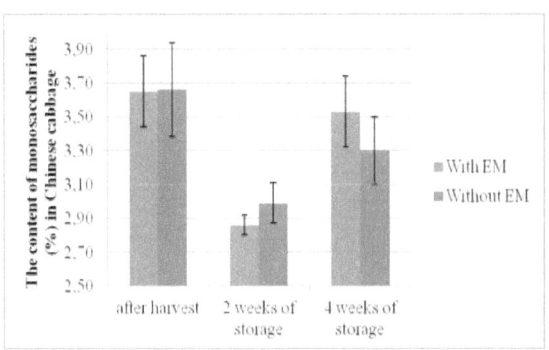

Figure 5. The content of monosaccharides (%) in Chinese cabbage (with and without EM); after harvest NS, 2 weeks of storage NS, 4 weeks of storage NS.

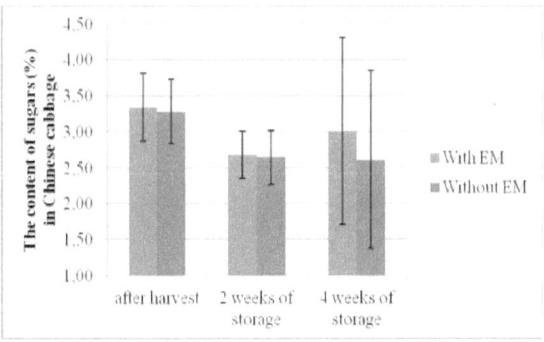

Figure 6. The content of sugars (%) in Chinese cabbage (with and without EM); after harvest NS, 2 weeks of storage NS, 4 weeks of storage NS.

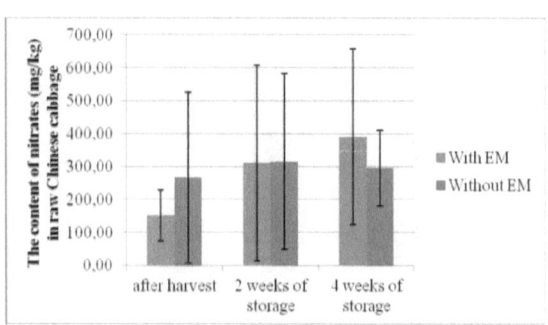

Figure 7. The content of nitrates (mg/kg) in raw Chinese cabbage (with and without EM); after harvest NS, 2 weeks of storage NS, 4 weeks of storage NS.

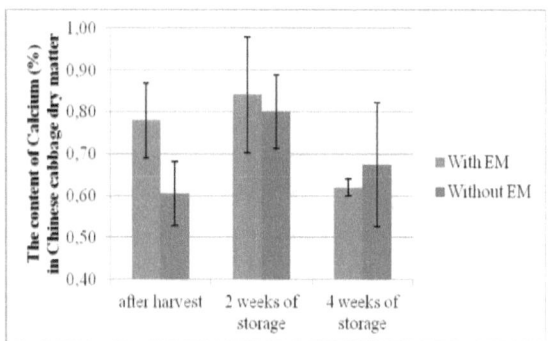

Figure 8. The content of calcium (%) in Chinese cabbage dry matter (with and without EM); after harvest 0.05%, 2 weeks of storage NS, 4 weeks of storage NS.

Paper 7

Influence of Effective Microorganisms on Soybean Seed Germination

Original paper reference, published in Estonian language:
Olle, M. (2014). Efektiivsete mikroorganismide mõju sojaoa seemnete idanevusele. In: Konverentsi ÄMahepõllumajanduse arengusuunad ï teadlaselt mahepõllumajanduseleñ toimetised: (Toim.) Metspalu, L., Luik, A.., 70 - 72.

Influence of Effective Microorganisms on Soybean Seed Germination

Abstract

As soybean has quite a long growing period, any effort made to shorten the growing period is beneficial. Therefore, it was important to conduct experiments to assess the effect of EM on soybean seed germination. The treatments included immersion of seeds for 60 minutes in activated EM solution 1:500 or in water (control). After immersion, the seeds were placed on moist filter paper in Petri dishes without drainage. There were significant differences among treatments in the germination percentage of soybean. The EM treatment showed the greatest number of germinated seeds.

Key words: Effective microorganisms, seed germination, soybean.

Introduction

Effective microorganisms (EM) technology was developed over 40 years ago by Dr. Tero Higa in Japan. EM are a mixed culture of beneficial, naturally-occurring micro-organisms, such as photosynthetic bacteria (e.g., *Rhodopseudomonas palustris*, *Rhodobacter sphaeroides*), lactobacilli (e.g., *Lactobacillus plantarum*, *L. casei*, and *Streptococcus lactis*), yeasts (e.g. *Saccharomyces* spp.), and Actinomycetes (*Streptomyces* spp.; Javaid, 2010). EM interacts with the soil-plant ecosystem to suppress plant pathogens and agents of disease, to solubilize minerals, to conserve energy, to maintain soil microbial-ecological balance, to increase photosynthetic efficiency, and fix biological nitrogen (Olle, 2013a; Subadiyasa, 1997).

As soybean has quite a long growing period, any effort made to shorten the growing period is beneficial. Giving the plant a good start may result in a healthier plant, earlier harvest and higher yield. Siqueira et al. (2012) stated that EM is a product of organic matter fermentation and its effects are similar to some biofertilizers.

Therefore, it was important to conduct experiments to assess the effect of EM on soybean seed germination.

Materials and Methods

The experiment was carried out in a greenhouse at the Jogeva Plant Breeding Institute in 2013. Seeds of the soybean variety SL were tested. Immersion of seeds for 60 minutes in activated EM solution 1:500 was compared with immersion in water (control) with 4 replications of 100 seeds each. The experiment was repeated. In the first experiment the seeds were treated on 15 January and the germinated seeds counted on 18, 19 and 20 January. In the second experiment, the seeds were treated on 21 January and the germinated seeds counted on 24, 25 and 26 January. The experiment was terminated when disease damage showed on the last day (in both experiments) of counting.

After immersion, the seeds were placed on moist filter paper in Petri dishes without drainage. The temperature was maintained at 20 ^{0}C during the day and 18 ^{0}C during the night. Water was applied whenever necessary. The smallest germ counted was at least 2 mm long.

Analyses of variance were carried out on the data obtained using Excel. Average, p value and standard deviation were calculated.

Results

There were significant differences between treatments in the germination percentage of the soybean seeds (Figure 1, 2).
The proportion of seeds that germinated was greater in the EM treated than in the control.

In the first experiment: three days after sowing, 49% more EM than control seeds had germinated, with the highest number of seeds germinated being 22. Four days after sowing, 40% more EM than control seeds had germinated, with the highest number of seeds germinated being 32.5. Five days after sowing, 36% more EM than control seeds had germinated with the highest number of seeds germinated being 48.75.

In the second experiment: three days after sowing, 51% more seeds had germinated in the EM than control, with the highest number of seeds germinated being 23.3. Four days after sowing, 39% more seeds had germinated in the EM than control, with the highest number of seeds germinated being 35.5. Five days after sowing, 44% more seeds had germinated in the EM than control, with the highest number of seeds germinated being 51.3.

Discussion

EM treatment improved soybean seed germination. Similar results with various vegetables have been found by Siqueira et al. (2012). Khan et al. (2011) postulated that low concentrations of EM speed up *Dalbergia sissoo* seed germination. Sangakkara and Attanayake (2013) found that EM 4 enhanced the germination of rice seeds. Mowa and Maass (2012) reported that EM enhanced *Harpagophytum procumbens* germination. Siqueira et al. (2012) stated also that EM is a product of organic matter fermentation and its effect is similar to some biofertilizers. This can be a reason why EM speeded up the germination process in our experiment.

Olle and Williams (2013) reviewed the beneficial effects of EM, amongst them enhancement of seed germination and suppression of plant pathogens and agents of disease. That can be the reason why seeds germinate better with EM than without.

Seed treatment with microbial inoculants such as EM may increase the rate of germination so that weakened seeds may survive to produce normal plantlets. Under field and greenhouse conditions, good quality seeds may not exhibit the beneficial effect of EM treatment; internal and external conditions may be favourable for germination regardless of treatment (Siqueira et al., 2012).

Sangakkara and Attanayake (2013) reported that microorganisms have been successfully used to promote seed germination in many crop species, and this concept could have an important role in agriculture. The results clearly present a protecting and growth-promoting role of these species in the germination process of many cereals. The causal mechanisms of this phenomenon were attributed to the protection of seeds from external stress factors and diseases during the germination process, when seeds and the emerging radicles and plumules are sensitive to the environment.

Several authors (Olle, 2013a, Olle, 2013b; Olle, Williams, 2013) have found that EM increases yield of most crops, so use of this product early in the season can give the soybean plant a goodstart, enhancing germination and increasing yield.

As soybean develops slowly and has quite a long growing period, it is clear to see the usefulness of EM in enhancing germination of this very important world crop species.

Conclusions

EM treatment of soybean seeds increased the proportion that germinated. As soybean develops slowly and has quite a long growing period, it is clear to see the usefulness of EM in enhancing germination of this very important world crop species.

References

Javaid, A. 2010. Beneficial Microorganisms for Sustainable Agriculture. - Sustainable Agriculture Reviews, 4, 347-369.

Khan, B. M., Hossain, M. K., Mridha, M. A. U. 2011. Nursery practice on seed germination and seedling growth of *Dalbergia sissoo* using beneficial microbial inoculants. Journal of Forestry Research, 2, 189-192.

Mowa, E., Maass, E. 2012. The effect of sulphuric acid and effective microorganisms on the seed germination of *Harpagophytum procumbens* (devil's claw). South African Journal of Botany, 83, 193-199.

Olle, M. 2013a. Efektiivsete mikroorganismide mõju kaalika saagile, keemilisele koostisele ja säilivusele. Agronoomia 2013, pages xxx-xxx, in press.

Olle, M. 2013b. Efektiivsete mikroorganismide mõju köögiviljade saagile, selle kvaliteedile ja säilivusele. Aiandusfoorum 2013, pages xxx-xxx, in press.

Olle, M., Williams, I. H. 2013. Effective microorganisms and their influence on vegetable production – a review. Journal of Horticultural Science & Biotechnology, xxx – xxx, in press.

Sangakkara, U. M., Attanayake, A. M. U. 2013. Effect of EM on Germination and Seedling Growth of Rice. http://www.emturkey.com.tr/TR/dosya/1-539/h/effect-of-em-on-germination-and-seedling-growth-of-rice.pdf (searched on 05.01.2013).

Siqueira, M. F. B., Sudré, C. P., Almeida, L. H., Pegorer, A. P. R., Akiba, F. 2012. Influence Of Effective Microorganisms on Seed Germination and Plantlet Vigor of Selected Crops. http://www.teraganix.com/category-s/1171.htm (searched on 12.12.2012).

Subadiyasa, N.N. 1997. Effective microorganisms (EM) technology: its potential and prospect in Indonesia. - Majalah Ilmiah Fakultas Pertanian Universitas Udayana, 16, 45-51.

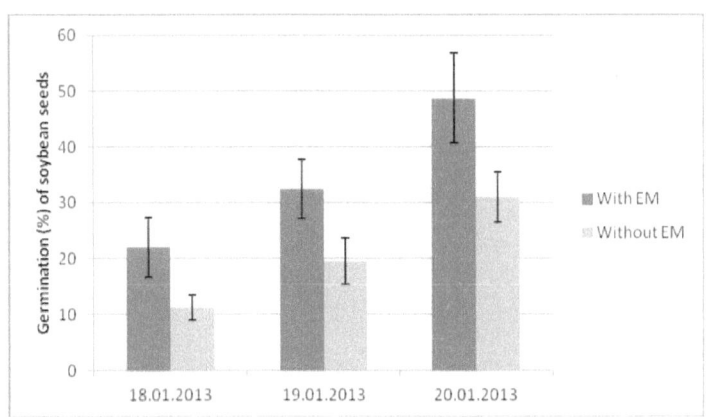

Figure 1. Germination (%) of soybean seeds 3, 4 and 5 days after immersion in EM solution or water alone in Experiment 1.

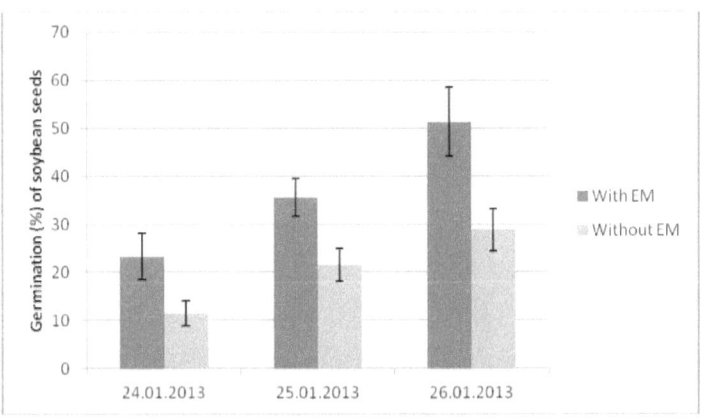

Figure 2. Germination (%) of soybean seeds 3, 4 and 5 days after immersion in EM solution or water alone in Experiment 2.

I want morebooks!

Buy your books fast and straightforward online - at one of the world's fastest growing online book stores! Environmentally sound due to Print-on-Demand technologies.

Buy your books online at
www.get-morebooks.com

Kaufen Sie Ihre Bücher schnell und unkompliziert online – auf einer der am schnellsten wachsenden Buchhandelsplattformen weltweit! Dank Print-On-Demand umwelt- und ressourcenschonend produziert.

Bücher schneller online kaufen
www.morebooks.de

OmniScriptum Marketing DEU GmbH
Heinrich-Böcking-Str. 6-8
D - 66121 Saarbrücken
Telefax: +49 681 93 81 567-9

info@omniscriptum.com
www.omniscriptum.com

www.ingramcontent.com/pod-product-compliance
Lightning Source LLC
Chambersburg PA
CBHW020455220526
45464CB00002B/995